RESORT HOTEL DESIGN
度假酒店设计

（澳）布伦顿·马里奥 编　李红 译

辽宁科学技术出版社
沈阳

图书在版编目 (CIP) 数据

度假酒店设计 / (澳) 布伦顿·马里奥编;李红译. —
沈阳:辽宁科学技术出版社, 2017.6
ISBN 978-7-5591-0129-7

Ⅰ.①度… Ⅱ.①布… ②李… Ⅲ.①饭店 – 建筑设
计 – 世界 – 图集 Ⅳ.① TU247.4–64

中国版本图书馆 CIP 数据核字 (2017) 第 072577 号

出版发行:辽宁科学技术出版社
　　　　　(地址:沈阳市和平区十一纬路 25 号　邮编:110003)
印 刷 者:辽宁新华印务有限公司
经 销 者:各地新华书店
幅面尺寸:245mm × 305mm
印　　张:52
插　　页:4
字　　数:260 千字
出版时间:2017 年 6 月第 1 版
印刷时间:2017 年 6 月第 1 次印刷
责任编辑:李　红
封面设计:李　莹
版式设计:李　莹
责任校对:周　文

书　　号:ISBN 978-7-5591-0129-7
定　　价:498.00 元

编辑电话:024-23280367
邮购热线:024-23284502
E-mail: 1207014086@qq.com
http://www.lnkj.com.cn

PREFACE
前言

DESIGNING ASPIRATIONAL SPA RESORTS
设计梦寐以求的SPA度假村

Spa design has come of age, surpassing the days of plain medically and clinically inclined white-washed walls and ceilings, around a massage table, more for kneading dough, than aimed at comfort and geared to ease away worries and stresses. Integrated spa resorts, whether boutique retreats, or chain brands, have become a holistic wellness experience and holiday escape. The body is not the only recipient, but also the mind, soul and spirit embrace the benefits of a fully comprehensive three-dimensional spa encounter. Audio, visual, scent, touch, texture and colour all make up the sensory journey one is invited to explore. Attention to detail is a paramount must – in short, everything.

The spa resort's design is the first introduction to the brand identity, products and services, followed swiftly and discreetly by the service staff or receptionist. The initial impression needs to be an impactful one. Warm, welcoming, relaxing, stress-free, any of these words and the like, need to be the ideal description of the ambience, at first glance. Nuance is made in the level of spa experience, quality of accommodation and service offering, as well as in the concept, design, materials and the style chosen.

We believe that the architecture and interiors set the scene and provide a pathway for the journey into the spa world. Creating the right atmosphere builds the brand, emphasises core values, puts guests immediately at ease and promotes customer loyalty. This ultimately enhances the success and profitability of the spa resort.

Once the resort design has set the scene, generated the mood and induced positive sentiments, it is then up to the staff, therapists, management, treatment menu and customer service offering, to complete the destination experience, and have clientele eagerly returning time and again. More than just a mere backdrop, spa resort design is an essential component in the quest for the perfect expression of the wellness experience.

SPA设计日趋成熟，已经超越了最初医院式简单而朴实的风格——以按摩台为中心，配以白色的墙与天花板，因为那时的SPA更加注重治疗按摩，而不是以舒适和减轻压力与烦恼为目的。综合性的SPA度假村，无论是精品度假胜地，还是连锁品牌度假酒店，都更注重于为客人带来全面的健康体验和轻松的度假经历。不仅让客人从身体上受益，而且要从思想、灵魂以及精神上全面享受三维SPA带来的轻松愉快之感。要从听觉、视觉、味觉、触觉、质感和色彩上，为客人带来一次充满探索的感官之旅。总之，要注意细节，细节决定一切。

客人对SPA度假村的品牌标识、产品和服务的认知是从SPA度假村的设计开始的，随后即是服务人员或接待人员细心周到的服务。最初的印象会给客人留下深刻的影响。温暖、热情、放松、释压，所有类似的词语都将是客人第一眼看到周围环境后最理想的描述。在SPA体验中，在食宿与服务质量上，以及在理念、设计、材料和风格的选择上，都要有细微玄妙的差别。

我们认为建筑与室内设计为通往水疗世界之旅提供了前提和通道。创造良好的氛围，建立品牌形象，强调核心价值观，让客人立即感到舒适与安逸，增强他们的信赖感。这些最终会成为SPA度假村成功与获益的根本。

首先由度假村的设计设定场景，愉悦了顾客的心情并且激发了积极的情绪，然后再由全体员工、治疗师、管理人员提供相应的治疗和服务，使顾客获得一次完整的度假体验，并且会让顾客渴望再度来此度假。SPA度假村的设计不仅仅是设计一个背景，更是追求健康体验的一个重要组成部分。

Brenton Mauriello　布伦顿·马里奥
Group CEO　团队执行总裁
dwp | design worldwide partnership　dwp|全球设计事务所

CONTENTS
目录

The Ritz-Carlton, Sanya

三亚丽思卡尔顿度假酒店

Completion date: 2008
Location: Sanya, China
Designer: Shanghai Institute of Architectural
Design & Research (Architect)
Chhada, Siembiedal Leung (Interior Design)
WATG Design (General Design)
Photographer: Chris Cypert
Site Area: 153,375sqm

竣工时间：2008年
项目地点：中国，三亚
设计师：上海建筑设计研究院（建筑师）
恰达思贝德梁设计公司（室内设计）
WATG设计公司（整体设计）
摄影师：克里斯·塞伯特
项目面积：153,375平方米

Classical architectural design inspired by Beijing's Summer Palace places the new Ritz-Carlton, Sanya resort among the most elegant in Asia. High pitched ceilings, dark rich wood, intricate carvings and mosaic detailing present a 'taste of tradition' amid modern 5-star comfort and facilities. The intent was to capture the timeless quality of traditional Chinese architecture, blending effortlessly with the natural environment. The focus of this design concept has been to create a five-star, world-class destination resort with international appeal.

At the main hotel entry, a large reflective pool flanked

gives way to a panoramic view of the entire resort and sea and sand beyond through an 'open window' design. There are also overtones of southern Chinese garden styles with whimsical sculpted lawns and lotus mazes among the resort's 150,000 square metre expanse. Throughout the resort, courtyards are dotted with water features, and outdoor decking areas create a further point of interaction with the water, be it sea or one of the resort's four pools, which includes a meandering lagoon.

The U-shaped design of the hotel's main building incorporates two wings housing 450 guest rooms including 334 oversized guest rooms at over 60 square metres, 67 rooms and suites on The Ritz-Carlton Club level with distinctive amenities, 16 suites and 33 private villas with individual plunge pools offering the ultimate privacy. Each comprises an oversized private balcony and marble-tiled five-fixture bathroom.

The Ritz-Carlton Club, which is the hotel within a hotel comprising 62 oceanfront view guest rooms and 5 suites including The Ritz-Carlton Suite, is located on the top level of the resort. The 33 one-, two- or three- bedroom villas with spacious private pools are all designed and decorated with novelty and unique

1. Zig Cigar Bar
2. Scene Lobby Bar
3. Sofia Italian Restaurant
4. Fresh 8 All-day Dining Restaurant
5. Pearl Chinese Restaurant
6. Sand Beach Grill Restaurant
7. Cube Pool Bar
8. Mood Spa Bistro
9. Recreation Room
10. Kids Club
11. Health Centre
12. Chi Studio
13. Spa Villa
14. Water Sports Centre
15. Tennis Court
16. Swimming Pool
17. Business Centre
18. Gift Shop
19. Uncle Martin's Secret Garden
20. Luxury Boutique
21. Ballroom and Function Rooms
22. Club Lounge
23. Ocean Pavilion

1. 雪阁——雪茄吧
2. 景台——大堂吧
3. 索菲娅——意大利餐厅
4. 鲜坊——全日餐厅
5. 润园——中餐厅
6. 海边——烧烤餐厅&酒吧
7. 池畔——泳池吧
8. 睦堂——水疗咖啡
9. 康乐中心
10. 儿童乐园
11. 健身房
12. 太极室
13. 水疗别墅
14. 水上活动中心
15. 网球场
16. 游泳池
17. 商务中心
18. 礼品商店
19. 马丁伯伯的小花园
20. 金茂时尚生活中心
21. 大宴会厅和多功能厅
22. 行政酒廊
23. 海景礼堂

taste, reminiscent of the picturesque water pavilions of China and the now popular water bungalows seen throughout South East Asia.

The sumptuous 2,788 square metre spa is a destination in its own right. Upon crossing an elegant stone bridge, guests embark upon a spa journey that is quintessentially ESPA. Beyond the welcome area lies an enchanting internal courtyard which features a Koi carp pond, lushly manicured landscapes, rock gardens and soothing water features designed to relax the mind and inspire the senses to enhance inner peace. ESPA has 24 private treatment rooms, four of which feature Italian made baths carved from stone. Couples seeking seclusion can book the ultimate spa experience in one of six specially appointed Private Spa Suites, which feature a private steam shower and luxurious outdoor Jacuzzi.

灵感来自北京颐和园的经典建筑设计，三亚丽思卡尔顿度假酒店坐拥亚洲旖旎风光。斜尖屋顶、深色实木材、复杂细致的雕刻和马赛克镶嵌装饰等，都展示了一个五星级酒店中的传统气息。该项目旨在打造耐久品质的中国传统建筑与自然环境的完美结合。该设计理念的核心是创建能够吸引世界目光的世界五星级度假酒店。

在酒店的主入口侧面有一个大的倒影池，通过一种"开窗"式的设计，可以将度假村与海洋、沙滩的美景尽收眼底。在这个150,000平方米的度假村里还有雕刻奇异的草坪和莲花池，体现了一种中国南方园林风格。整个度假村的庭院都点缀着水文景观，户外露台区更与水紧密结合，使之与海相接或与度假村中四大泳池相接，其中包括一条蜿蜒的环礁湖。

酒店主建筑为U形设计，两翼设有450间客房，其中包括334间面积超过60平方米的景致客房，67间位于行政楼层的特色客房和套房，16间观景套房以及33个拥有独立泳池的私家别墅。每间客房都设有一个特大的私人阳台和大理石铺砌的五项卫浴组合浴室。

丽思卡尔顿行政楼层——酒店中的酒店——位于度假酒店的顶层，拥有62间豪华海景客房及包括丽思卡尔顿总统套房在内的5间套房。33座私家泳池别墅分别为一房、两房和三房格局，设计新颖，装潢考究，尽显独特品位，令人联想到中国独特的水亭和现在东南亚盛行的水上屋。

2788平方米的豪华水疗中心自成一体。穿过典雅的石桥，客人们由此进入ESPA精粹之旅。接待处旁别具一格的水景花园、鲤鱼池塘、精心修剪的花草以及平缓的流水均是为了让客人放松情绪、激活身体感官以达到内心的安宁而专门设计的。ESPA可提供24个私人理疗室，其中4间配有石纹雕刻的意大利浴缸。6间为配有单独蒸汽室和户外浴缸的SPA套房，专为寻找私密SPA极致体验的情侣们而设。

1. Resort exterior at night
   酒店外观夜景
2. Pool villa at night
   泳池别墅夜景
3. Villa sunset path
   夕阳下的别墅小径
4. Resort plan
   度假村平面图
5. Scene A restaurant in the centre of the hotel
   景台——位于酒店中心位置的餐厅

6. Shui Man meeting room
   水满董事会议厅
7. Sofia Italy Food Restaurant interior overall
   索菲娅——意大利美食餐厅内全景
8. Sofia Teresa private dining room
   索菲娅——特丽莎私人餐室
9. Zig-Cigar Bar
   雪阁——雪茄吧

10. Espa exterior
   Espa外观
11. Espa tea lounge overview
   Espa茶歇室全景
12. Espa reception
   Espa前台

The St. Regis Saadiyat Island Resort

萨迪亚特岛瑞吉度假酒店

Completion date: 2011
Location: Abu Dhabi, United Arab Emirates
Designer: Creative Kingdom (Conceptual Architect),
Woods Bagot (Architects),
Hirsch Bedner Associates (Interior Design).
Photographer: The St. Regis Saadiyat Island Resort

竣工时间：2011年
项目地点：阿拉伯联合酋长国，阿布扎比
设计师：创新王国股份有限公司（概念设计）
伍兹贝格建筑事务所（建筑设计）
赫希贝德纳联合酒店顾问有限公司（室内设计）
摄影师：萨迪亚特岛瑞吉度假酒店

The St. Regis Saadiyat Island Resort is framed by the prestigious Saadiyat Beach Golf Course, and a nine-kilometre pristine beach. A tree-lined boulevard leads the visitor to a roundabout filled with exotic plants, and from here you find the exclusive main entrance, which comes complete with a stylish marquee and a water fountain. Views to the stunning Saadiyat Beach Golf Course and the beach are fully exploited. On the opposite side of the site, there is a separate entrance for the conference centre and meeting room facilities. The hotel's air of mystique is created by thoughtful Mediterranean architecture and a contemporary

interior design, which is further enhanced by an intelligent use of natural products and elements. This calming ambience flows through the food and beverage outlets that provide sumptuous meals and refreshing drinks all the way to one of the largest function facilities in the region.

The hotel offers 377 rooms and suites, all with private balconies, offering spectacular ocean views, 3,000 sqm ballroom, complemented with four meeting rooms and a boardroom, 6,000 sqm of retail space encompassing restaurants, lounges and boutiques, St. Regis Athletic Club 3,500 sqm, the region's first Iridium Spa with 12

treatment rooms and 3,500 sqm of space, Sandcastle Club – a children's club featuring indoor and outdoor activities, seven distinctive food & beverage venues, over 2,000 parking spaces, two tennis courts and two squash courts, five swimming pools including a 25-metre indoor lap pool and an adult-only pool.

The 377 guest rooms and suites range from Superior Rooms, at 55 sqm with a 10 sqm balcony, to Suites from 83 sqm with a 10 sqm balcony, to the magnificent Royal Suite, at 982 sqm with a spacious 179 sqm balcony. The resort boasts lavish yet intimate spaces with Superior Rooms, St. Regis Suites and

1. St. Regis Hotel   3. St. Regis Apartments   1. 瑞吉酒店   3. 瑞吉公寓式客房
2. The Regal Ballroom   4. St. Regis Villas   2. 豪华宴会厅   4. 瑞吉别墅

magnificent Royal Suite overlooking either Saadiyat's pristine beach or the impressive Saadiyat Beach Golf Club.

The Iridium Spa is rare and refined. From arrival to departure, Iridium Spa provides guests with the rarest of luxuries: time. It is here where guests can enjoy exclusive access to one of the world's most coveted benefits.

萨迪亚特岛瑞吉度假酒店由著名的萨迪亚特海滨高尔夫球场和长达9千米的原始海滩包围着。一条林荫大道通往环岛，岛内种满各种奇异植物。从环岛出发就来到了酒店唯一的主入口，主入口配有气派的遮阴篷和喷泉。萨迪亚特海滨高尔夫球场和海滩的美丽景色尽收眼底。酒店另一侧有一个单独入口通往会议中心和其他会议室。

酒店地中海风格的建筑和当代室内设计风格给酒店带来神秘的气息，各种天然产品与自然元素的合理使用使这种神秘气息更加浓重。这种宁静的氛围从提供美味佳肴与清凉饮品的餐饮区蔓延至该区最大的会议功能厅。

酒店提供377间客房与套房，每间均带有私人阳台并可以欣赏美妙海景；3000平方米的宴会厅，配有4间会议室与一间董事会议室；6000平方米的零售区包括餐厅、休闲区和精品店；3500平方米的瑞吉体育俱乐部；该地区第一家铱SPA，拥有12间治疗室，占地3500平方米；沙堡俱乐部——一家以室内外活动为特色的儿童俱乐部；7个独具特色的餐饮区；2000多个停车位；两个网球场和两个壁球场；5个游泳池，包括一个25米长的室内小型健身游泳池和一个成人泳池。

377间客房和套房不仅包括55平方米高级客房（阳台10平方米），面积不小于83平方米的套房（阳台10平方米），还包括面积达982平方米、装饰华丽的皇家套房（配有179平方米宽敞阳台）。该酒店设有奢华而又私密的高级客房——瑞吉套房和华丽的皇家套房——可以俯瞰萨迪亚特原始海滩或者萨迪亚特海滨高尔夫俱乐部。

铱SPA精致而稀有。自始至终，铱SPA为客人提供的是最稀有的奢侈品：时间。客人们在此享用的是世人垂涎的珍稀资源，获得的是独一无二的尊贵体验。

The
DRAWING ROOM
St Regis · Saadiyat Island

9. The Iridium Spa treatment room
   铱SPA治疗室

10. The Iridium Spa corridor
    铱SPA走廊

11. The Iridium Spa pool area
    铱SPA泳池

12. The Iridium Spa reception
    铱SPA前台

13. The Iridium Spa treatment room
    铱SPA治疗室

14. The Iridium Spa treatment room and pool
铱SPA治疗室与泳池

15. Superior Room
高级客房

16. St. Regis Suite
瑞吉套房

17. Ocean Suite
海景套房

18. Superior Sea View Room
高级海景客房

19. Ocean Suite bathroom
海景套房浴室

20. Bathroom in Superior Sea View Room
高级海景客房浴室

Hilton Bodrum Turkbuku Resort & Spa

图克布库希尔顿博德鲁姆水疗度假酒店

Completion date: 2011
Location: Bodrum, Turkey
Designer: OdaYapi
Photographer: Reklamaks
Area: 115,000sqm

竣工时间：2011年
项目地点：土耳其，博德鲁姆
设计师：OdaYapi
摄影师：Reklamaks摄影工作室
项目面积：115,000平方米

Hilton Bodrum Turkbuku Resort & Spa, nestled on the North end of the historical peninsula, welcomes its guests to extraordinary comforts amidst the green hills of Cennet Koyu (Paradise Bay) and the natural wonder of its private turquoise blue cove. Life at Hilton Bodrum Turkbuku Resort & Spa is an embracing awareness experience starting on your first step into the lobby. This exclusive quality is felt in each and every detail of the interior design, the array of accommodation choices, the terraces reaching to embrace the Aegean waters and, last but not least, the trusted services of the Hilton tradition. This paradise, reconstructed on 1,150,000 square metres,

presents all the traditional marks of Hilton's personalised service standards. The Resort's 433 guestrooms, 53 spacious suites and 8 detached family villas feature the most luxurious details of an architectural style based on comfort and aesthetics.

Hilton Bodrum Turkbuku Resort & Spa presents the largest and best-equipped meeting and convention options in the Bodrum area with its 1,200 square metres capacity. The specially designed a 6-metre high ballroom, with a capacity of 1,500 delegates, which can be divided into three meeting rooms of 500 square metres. The 650-square-metre, sunlit and sea-view foyer can be utilised for exhibitions, coffee breaks, cocktail parties and gala dinners.

Meaning 'health through water' in Latin, Sanus Per Aquam, better-known as SPA, takes on an entirely new meaning at Hilton Bodrum Turkbuku. At this resort located in the heart of a perfect sea, special treatments from around the world are offered as individual tailor-made treatment packages to counter stress and improve health and appearance. Its fitness centre features 16 different therapy rooms, saunas, steam baths, Turkish bath, solarium, Jacuzzi, indoor fresh water pool, indoor Thalasso pool and other special pools.

1. Villas	9. Amphitheatre	1. 别墅区	9. 圆形剧场
2. Villa platform	10. Rooms	2. 别墅站台	10. 客房区
3. Massage	11. Tennis court	3. 按摩区	11. 网球场
4. Beach club	12. Entrance lobby	4. 海滨俱乐部	12. 入口大厅
5. Café Turc & Tea House	13. Ala Mer Bar	5. 土耳其咖啡厅与茶馆	13. Ala Mer酒吧
6. Pool bar	14. Seaport Restaurant	6. 泳池吧	14. 海港餐厅
7. Water slides	15. Paparazzi Restaurant	7. 水滑梯	15. Paparazzi餐厅
8. Kids club	16. Water sport	8. 儿童俱乐部	16. 水上运动区

图克布库希尔顿博德鲁姆水疗度假酒店地处历史悠久的博德鲁姆半岛的北端，坐落在天堂湾翠绿群山的怀抱之中，坐拥土耳其碧绿峡谷这一天然奇观。该度假酒店以其舒适的环境和贴心的服务诚邀海内外广大游客。游客一踏入酒店大堂，一种舒适和宽慰的感觉油然而生。无论是各式客房的设计，还是拥抱爱琴海的宽大露台，或是希尔顿传统的可信服务等，酒店内部的每一处细节设计都彰显出了别致与舒适。

图克布库希尔顿博德鲁姆水疗度假酒店经重建后，变成了一个面积达1,150,000平方米的度假天堂。这里提供给游客希尔顿式私人服务标准的传统酒店式服务，令游客备感亲切。该度假村共有433间客房、53间宽敞的套房和8个独立的家庭别墅，所有房间的设计都在舒适和唯美的基础上体现了无处不在的奢华。

图克布库希尔顿博德鲁姆水疗度假酒店拥有博德鲁姆地区面积最大、设施配备最齐全的会议室。会议室面积达1,200平方米。采用特殊手法设计的宴会厅，高6米，被分成了3个面积达500平方米的会议室。面积达650平方米的海景门厅，终日阳光普照，可供客人举行展览、享受咖啡的休闲时光、举行鸡尾酒会或其他一些晚宴等。

被普遍称作SPA的"Sanus Per Aquam"，意思是水疗，这种热水疗法在图克布库希尔顿博德鲁姆水疗度假酒店被赋予了全新的含义。这个坐落在完美的大海中央的度假村为每位游客量身打造，提供一系列国际标准的健康疗法，为游客消除疲劳，改善其身体状况，并提供美容护理服务。健身中心拥有16个独具特色的疗养室、桑拿室、蒸汽浴室、土耳其浴室、日光浴室、极可意水流按摩浴缸、室内淡水游泳池、室内海水浴池和其他特制泳池等。

1. Lobby
   酒店大堂

2. Resort exterior view
   度假村外景

3、4. Resort beach view
   度假村海滩景色

5. Resort plan
   度假村平面图

6. Lobby entrance
   大堂入口

7. Lobby terrace
   大堂露台

8. Main restaurant
   主餐厅

9. Sultan Sofrasi Restaurant
   苏丹索福瑞餐厅

10. Outdoor massage area
    户外按摩区
11. Spa entrance
    SPA入口
12. Turkish Bath
    土耳其浴
13. Massage area
    按摩区

14. One-bedroom suite living room
一居室套房客厅

15. Presidential suite bedroom
总统套房卧室

16. One-bedroom suite bedroom
一居室套房卧室

17. Presidential suite living room
总统套房客厅

18. Turkish Bath
土耳其浴室

Kempinski Hotel Barbaros Bay

凯宾斯基巴巴罗斯湾酒店

Completion date: 2005
Location: Bodrum, Mugla , Turkey
Designer: Cengiz Eren
Photographer: Adrian Houston
Area: 40,000sqm (Outdoor areas),
70,000sqm(Indoor areas)

竣工时间：2005年
项目地点：土耳其，穆拉，博德鲁姆
设计师： Cengiz Eren
摄影师：艾德里安·休斯顿
项目面积：40,000平方米（室外区域），70,000平方米（室内区域）

Located on a pristine bay of the Bodrum Peninsula, the 5 star luxurious hide-away resort Kempinski Hotel Barbaros Bay is home to 24 lavish suites, 149 luxurious rooms and 36 serviced residences each with its own balcony and generous views of the Aegean Sea. Besides its 5,500sqm award winning Spa, Kempinski Hotel Barbaros Bay offers a world of leisure experiences with a private bay, private sandy beach, a massive infinity pool, private yacht slips, a private helipad, a Jacuzzi terrace, outdoor and indoor conference and banqueting venues, pirates of Barbaross kids club, an extensive library, five restaurants offering tastes from around the globe, and

two bars.

The placid blue waters of the Kempinski Hotel Barbaros Bay's infinity pool merge with the turquoise Aegean, stretching out to the horizon where the terracotta orb of the sun sinks into the waters of the Aegean and the pool painting them deep and enchanting shades of purple. As the ultimate laid-back setting for casual events, overlooking stunning views of the Aegean, the Chill-out Lounge can seat 110 or accommodate up to 160 guests for a reception.

As expansive and luxuriously tranquil as their limitless views of the blue Aegean, the 201-square-metre Presidential Suites each have their own design language – Asian Flavour or Ottoman Desire, and both feature expansive terraces with al fresco whirlpools, as well as a host of exclusive amenities and private services. Each of the one-bedroom suites is the perfect couples retreat, featuring a large living room, a powder room, a master bath with Jacuzzi, and a bedroom with a balcony or terrace that is the perfect place to greet the day, commanding a panoramic view of the Aegean. After a demanding day of sunbathing, swimming and sauna, the en suite living area of the Junior Suites is the perfect place to contemplate life's most challenging decisions, like which of the gourmet restaurants to choose for dinner…

01. First aid room	1. 急救室
02. Reception lobby	2. 接待大堂
03. Concierge	3. 门房
04. Business centre	4. 商务中心
05. Shops	5. 商店
06. Gazebo lounge & terrace	6. 露台休闲吧与阳台
07. Library	7. 图书室
08. Olives Restaurant	8. 橄榄餐厅
09. Barblue	9. 百兰
10. Kempi kids	10. 凯宾斯基儿童乐园
11. Game room	11. 游戏室
12. Chill-out	12. 放松室
13. Meeting rooms	13. 会议室
14. Pool arena	14. 泳池竞技场
15. Pool bar & restaurant	15. 泳池酒吧与餐厅
16. Spa & gym	16. 水疗中心与运动中心
17. Saigon club	17. 西贡俱乐部
18. La Luce	18. 电灯
19. Marina	19. 散步道
20. Sundeck	20. 日光浴处所
21. Kids' playfield	21. 儿童运动场
22. Barbarossa beach & grill	22. 巴巴罗萨
23. Water sports	23. 水上运动区
24. Volleyball-Bastketball-Soccer	24. 排球－篮球－足球场
25. Marina stone area	25. 滨石地区

凯宾斯基巴巴罗斯湾酒店坐落于博德鲁姆半岛的一个天然海湾上，是一家五星级的度假胜地。凯宾斯基巴巴罗斯湾酒店拥有24间豪华套房、149间豪华客房和36栋服务公寓，每个房间都带有一个独立的阳台，并可一览爱琴海的全景。除了曾获奖的水疗中心之外，凯宾斯基巴巴罗斯湾酒店还提供诸多休闲体验，拥有1个私人港湾、1个私人沙滩、1个巨大的无边际泳池、私人快艇、私人直升机停机坪、1个放置按摩浴缸的露台、各种室内外会议室及宴会厅、巴巴罗斯儿童俱乐部、1个宽敞的图书馆、提供世界各种美食的5间餐厅及2个酒吧。

凯宾斯基巴巴罗斯湾酒店无边际泳池中宁静而湛蓝的池水与蓝绿色的爱琴海融为一体，天水相接，海天一色，爱琴海的海水与泳池的池水将太阳这个赤红色的球体淹没，池水用浓墨重彩将天空和太阳加以渲染，烘托出一种紫色爱恋。Chill-out Lounge宴会厅拥有110个座位，可接待160名客人，餐厅中可一览爱琴海的美丽景色，是休闲娱乐的顶级聚会场所。

面积达201平方米的总统套房如同蓝色爱琴海一般宁静、广阔、大气、无边无际。每间套房都有其独特的设计语言——亚洲风韵或土耳其梦想，两间总统套房都拥有广阔的阳台，阳台上摆放着室外涡流浴缸，并配备多种专用设施，提供多种私人服务项目。一居室套房是情侣度假的最佳选择，每间套房都拥有1个宽敞的客厅、1个化妆室、1个配备极可意水流按摩浴缸的主浴室和1间带阳台的卧室，阳台是观赏日出景色的最佳地点，并可一览爱琴海的全景。体验完日光浴、游泳和桑拿等活动项目之后，游客可以选择在标准套房的客厅中思考生活中一些棘手的问题。正如美食餐厅是最理想的晚宴地点一样，这里是最佳的沉思地点。

1. Resort exterior
   度假村外观
2. Resort plan
   度假村平面图
3. Private dining terrace
   私人餐饮露台
4. Saigon Club terrace
   西贡俱乐部露台
5. Treatment terrace
   露台治疗区
6. Spa Chill Out
   水疗放松区
7. Corridor
   通道
8. Spa lounge area
   水疗休闲区

9. Spa treatment room
   水疗中心治疗室
10. Waysu Pool
   慰舒泳池
11. Spa treatment room
   水疗中心治疗室
12. Spa Suite wet area
   水疗套房盥洗室
13. Hamam
   哈曼浴室
14. Kempinski hotel guest room
   凯宾斯基酒店客房
15. Kempinski hotel guest room
   凯宾斯基酒店客房
16. Kempinski hotel guest room
   凯宾斯基酒店客房
17. Suite Jacuzzi
   套房内极可意水流按摩浴缸

Capella Singapore

新加坡嘉佩乐酒店

Completion date: 2009
Location: Singapore
Architect: Foster + Partners
Interior Design: Jaya International Design
Area: 121,400sqm (about 40% is built up. The remaining 60% of the land is natural greenery.)

竣工时间：2009年
项目地点：新加坡
建筑师：福斯特建筑事务所
室内设计：Jaya国际设计
项目面积：121,400 平方米（40%已建成；其余的60%为自然景观）

Located on Singapore's premier resort destination, Sentosa Island, as the flagship property for Capella Hotels and Resorts in Asia, Capella Singapore offers an inspiring natural setting while providing easy access to Singapore's CBD. This engaging duality is echoed in the hotel's design, with a masterfully restored colonial building forming a centre-piece of modern architectural elements and sculpture gardens rich with contemporary art.

Capella Singapore promises the ultimate in personalised service and represents a new standard of luxury in Asia, combining the best of old and new Singapore. The focal point of Capella Singapore is Tanah Merah, a colonial

structure built in the 1880s by the British Military to host their galas. The two-storey bungalow features colonial white columns, open-air verandahs and a red-tile roof, complemented by a new, contemporary extension, designed by Foster + Partners, London.

Capella Singapore offers the most spacious accommodations in Singapore. The 112 guest rooms include 61 premier guest rooms, 11 suites, 38 villas featuring private outdoor showers and bathtubs, and two Colonial Manors. In addition, Capella Singapore offers the opportunity to live in the hotel for as long as the guests wish to, with full access to the hotel's facilities. These long-term offerings will include 62 sea-facing suites, 10 three-room penthouses and 9 manors with private pools.

Auriga, a compelling new spa brand, has made its Southeast Asia debut at Capella Singapore. Named after the constellation whose brightest star is Capella, Auriga provides guests with a new wellness philosophy based on the phases of the moon. Auriga at Capella Singapore offers over 1,115 sqm of contemporary Asian space, featuring nine experience rooms with private gardens for a distinct indoor/outdoor experience. The spa also features a vitality pool with an ice fountain, a herbal steam bath and two experience showers.

新加坡嘉佩乐酒店坐落在新加坡顶尖度假景点——圣淘沙岛（嘉佩乐酒店和度假村的旗舰产业），提供灵动的自然美景和便利的城市交通（便于往返新加坡城区）。这两种迷人的元素在酒店设计中得到了体现，经过巧妙修复的殖民地风格建筑内洋溢着现代元素，雕塑花园里也摆满了现代艺术品。

新加坡嘉佩乐酒店承诺终极的个性化服务，呈现了亚洲式奢华的新标准，将新加坡的新旧元素完美地结合在一起。嘉佩乐酒店度假村的焦点是丹娜美拉——一座建于19世纪80年代由英国军队建造殖民地风格建筑。两层高的小楼以极具殖民地色彩的白色柱子、露天游廊和红砖屋顶为特色，而伦敦的福斯特建筑事务所则为建筑增添了全新的现代元素。

新加坡嘉佩乐酒店提供新加坡最宽敞的住宿。112套客房中包括61套顶级客房、11套套房、38座别墅（以私人露天淋浴和浴缸为特色）和两座殖民特色庄园。此外，嘉佩乐酒店还提供长期服务，宾客能够随时使用酒店设施。这些长期的服务设施包括62套朝海套房、10套三室顶层公寓和9座带有私人泳池的庄园。

Auriga是新加坡嘉佩乐的招牌水疗及保健中心，于东南亚地区首次登场，Auriga的中文意思是御夫座，嘉佩乐就是御夫座中最闪耀的星星。而Auriga芳疗中心则以月亮的周期为基础为宾客带来了全新的健康哲学。芳疗中心提供了1,115平方米的现代亚洲空间，以9间配有私人花园的体验室为特色，提供独具特色的室内外交替体验。同时，芳疗中心的活力泳池还配有冰喷泉、草药蒸汽浴和两个体验淋浴间。

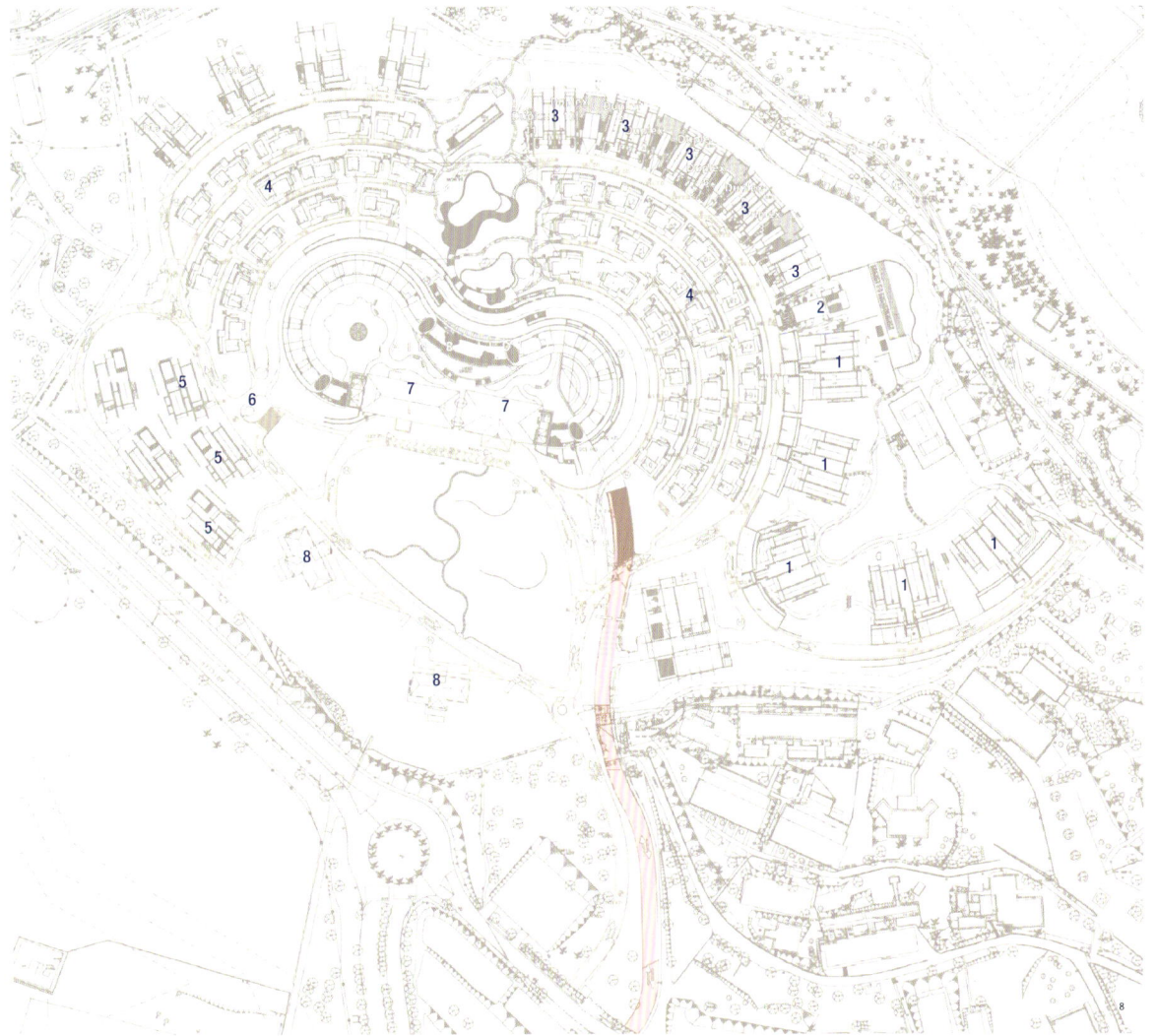

1. Suite	5. Manor villas
2. Clubhouse	6. Guard house
3. Duplex	7. Main building
4. Garden villas	8. Colonial villas

1. 套房	5. 庄园别墅
2. 俱乐部会所	6. 警卫室
3. 复式套房	7. 主楼
4. 花园别墅	8. 殖民别墅

1. Resort exterior
   度假村外观
2. Resort night view
   度假村夜景
3. Building facade and landscape
   建筑外观与景观
4. Resort Courtyard Daybreak
   黎明时分度假村庭院
5. Bob's Bar terrace
   Bob's酒吧露台
6. Resort night view
   度假村夜景
7. Bob's Bar at night
   Bob's酒吧夜景
8. Resort plan
   度假村平面图
9. Resort exterior sculpture
   度假村雕塑

10. Cassia Restaurant
    凯嘉餐厅
11. Bob's Bar
    Bob's 酒吧
12. Ballroom
    宴会厅
13. Sentosa boardroom
    圣淘沙会议室
14. Spa entrance
    芳疗中心入口
15. Spa reception
    芳疗中心接待处
16. Spa recovery area
    芳疗中心恢复区
17. Spa couple treatment room
    芳疗中心双人理疗室

18. Colonial Manor dining room
    殖民风格庄园餐厅
19. Bliss room
    新人专属更衣套房
20. Resort reception
    度假村接待处
21. Premiere room plan
    至尊套房平面图

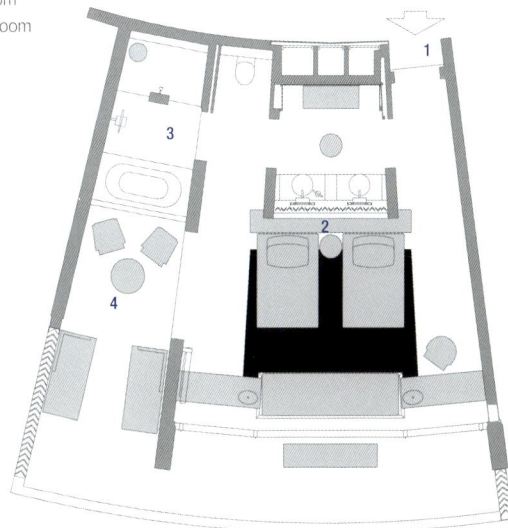

1. Entrance lobby
2. Bedroom
3. Bathroom
4. Living room

1. 门厅
2. 卧室
3. 浴室
4. 起居室

24

27

25

28

26

29

22. Colonial Manor bedroom
    殖民风格庄园卧室

23. Tea lounge
    茶吧

24. Sentosa Suite
    圣淘沙套房

25. Premier room
    至尊客房

26. Capella Suite
    嘉佩乐套房

27. Premiere seaview room
    至尊海景房

28. Simplex Suite
    单层套房

29. Manor kitchen
    庄园套房厨房

Sofitel Philippine Plaza

索菲特广场酒店

Completion date: 2010
Location: Manila, Philippines
Designer: National Artist for Architecture,
Leandro Locsin, Ildefonso P. Santos, Spin Design Studio
Photographer: Sofitel Philippine Plaza
Area: 69,800sqm (floor area), 59,000sqm (total land area)

竣工时间：2010年
项目地点：菲律宾，马尼拉
设计师：莱昂德罗·洛克辛（国家级艺术家）、伊尔德丰
索·P·圣多斯，Spin Design Studio设计事务所
项目面积：69,800平方米（建筑面积），59,000平方米（总面积）

Boasting of spectacular views of the famous Manila Bay, and lush landscapes and greenery, the exquisite tropical setting within the city is unrivalled. Coupled with bright, open spaces, and elegant, minimalist interiors with a touch of traditional Filipino décor, Sofitel Philippine Plaza is a gem in the bustling city.

The hotel has a total of 609 rooms, including 2 Opera Suites, 4 Luxury Suites, 8 Prestige Suites, 38 Sofitel Suites and the magnificent 398 square metre Imperial Suite. The Imperial Suite's exquisitely appointed rooms – all three bedrooms, foyer, living, dining and music rooms, along with a library, a kitchen and a service bedroom, are fit for royalty and celebrity.

At the Luxury Club Sofitel Room, total comfort and rejuvenation is the promise of the award-winning MyBed, designed exclusively for Sofitel guests. With the spacious 42 sqm guest rooms, the experience of a blissful rest is beyond compare. The Superior Room has stylishly minimalist interiors showcasing the finest in Filipino craftsmanship, featuring exquisite capiz shells and coconut inlays.

With elegant interiors of modern natural elements such as Philippine shells, dark wood and natural accents, LeSpa is a design oasis for the soul. Its modern facilities ensure the smoothest, most luxurious spa experience. A lounge, dry and wet sauna, foot spa area, manicure and pedicure stations, nine treatment suites that include a duo suite with huge bath tub, and spa suite with a state-of-the-art Trautwein Crystal Bath and sundeck, all make for an unforgettable stay at LeSpa.

The hotel's signature facility, however, is the beautiful lagoon-shaped swimming pool with two giant slides and cascading waterfalls that make it a breathtaking centerpiece in the midst of landscaped tropical gardens dotted with palm trees. Getaways at Sofitel Philippine Plaza are designed to be truly memorable, as its French savoir faire and gentle Filipino hospitality continue to delight and disarm.

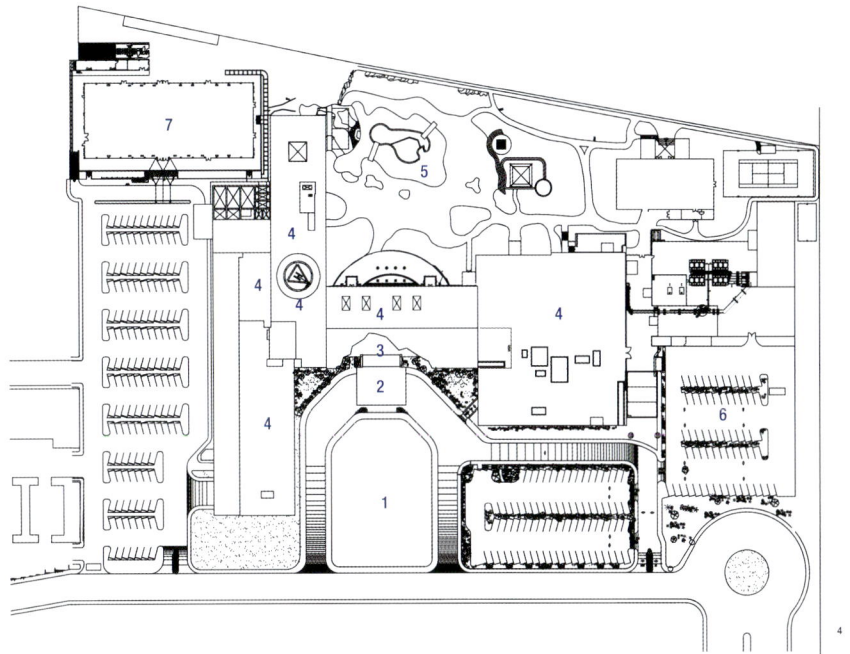

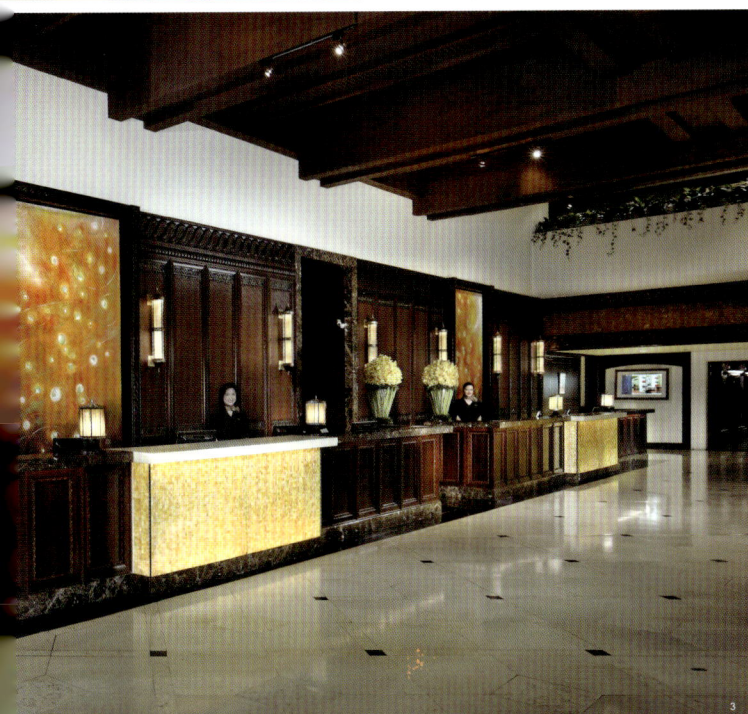

1. Front lawn　　　　　　　1. 前厅草坪
2. Main entrance canopy　　2. 主入口华盖
3. Main lobby　　　　　　　3. 大堂
4. Roof deck　　　　　　　4. 屋顶板
5. Swimming pool　　　　　5. 游泳池
6. Parking　　　　　　　　6. 停车场
7. Harbour garden　　　　　7. 港湾花园

索菲特广场酒店坐拥马尼拉湾久负盛名的壮观景色，周围树木环绕，苍翠繁茂，优美的热带景观在菲律宾堪称首屈一指。索菲特广场酒店拥有明亮宽敞的空间和极简抽象派艺术风格的室内设计，同时散发着菲律宾装饰的传统韵味，优雅精致，是繁华都市中的一块瑰宝。

该酒店共有609间客房，其中包括2间歌剧院套房、4间豪华套房、8间总统套房、38间索菲特套房和壮观的皇室套房。皇室套房面积达398平方米，拥有3间特殊打造的精致卧室、1个门厅、1个客厅、1个餐厅和1间音乐室，还有1个图书馆、1间厨房和服务型卧室，非常适合皇室贵族居住，也可举行一些庆祝活动。

奢华的索菲特俱乐部客房专为客人打造了曾获大奖的MyBed卧床，让游客享受无比的舒适和放松。42平方米的宽敞卧室带给游客无可比拟的至尊享受。高级客房内采用极简主义风格的时尚装饰，展示了菲律宾最佳的手工艺技术，室内主要以精致的贝壳和椰子镶嵌物为特色。

LeSpa水疗中心采用菲律宾贝壳、黑木和天然装饰物等现代自然元素来装点优雅的室内环境，打造了一个灵魂放松的绿洲。现代化的水疗设施为游客带来一次无比舒畅、无比奢华的水疗体验。水疗中心配备宽敞的休息室、干湿蒸两用桑拿房、足浴区、美甲区与足疗区、9间理疗套房，理疗套房内还有一间双人套房，套房带有一个巨大的浴盆，还有一个水疗套房，套房内配备了先进的Trautwein水晶浴室和阳台，配套设施应有尽有，令客人享受到难忘的LeSpa体验。

然而，菲律宾广场酒店的招牌设施要属环礁湖形状的、美丽的游泳池。游泳池两旁安装了巨大的滑梯，还有飞流直下的瀑布，形成了棕榈树点缀的热带景观园林中最别致、最壮观的景色。

菲律宾广场酒店的Getaways 体验令游客流连忘返，这里结合了法国的设计技艺与菲律宾热情好客的待客之道，令人心旷神怡、身心得到彻底放松。

1. Resort exterior landscape at dusk
   度假村黄昏外景
2. Building facade
   建筑外观
3. Lobby reception
   大堂接待处
4. Resort plan
   度假村平面图
5. Ballroom
   宴会厅
6. Restaurant
   餐厅
7、8. Fever Luxe Lounge
   奢华发烧酒吧
9. Ballroom
   宴会厅
10. Conference centre
   会议中心
11. Club Millime lounge
   米利姆俱乐部酒吧
12. Meeting room
   会议室
13、14. Club Millime exterior lounge
   米利姆俱乐部户外休闲吧

15. Spa reception and lounge
    水疗中心接待处与休闲吧
16. Spa treatment room
    水疗中心治疗室
17. Spa entrance
    水疗中心入口
18. Spa corridor
    水疗中心走廊
19、20. Spa treatment room
    水疗中心治疗室

21. Suite living room
    套房客厅
22. Suite bedroom
    套房卧室
23~25. Suite living room
    套房客厅

dusitD2 Baraquda Pattaya

芭提雅都喜D2芭拉古达酒店

Completion date: 2008
Location: Pattaya, Thailand
Designer: dwp / design worldwide partnership
Photographer: dusitD2 Baraquda
Area: 10,000sqm

竣工时间：2008年
项目地点：芭提雅，泰国
设计师：dwp / 世界设计合作公司
摄影师：芭提雅都喜D2芭拉古达酒店
项目面积：10,000平方米

Situated in one of the most vibrant and dynamic locations in the seaside city of Pattaya, the dusitD2 Baraquda hotel targets clients looking for a different holiday experience in Southeast Asia. The hotel features luxurious guest rooms, dynamic meeting spaces, delicious restaurants, the cool Deep Bar, a rooftop lounge with stunning sunset views, an exclusive spa and a spectacular poolside catwalk, as well as a host of other convenient facilities and amenities.

The initial concept briefing from the owner was to create an adventurous yet relaxing environment,

symbolic of the silver metallic sheen of barracuda fish. The design was challenging as the site is long and narrow with minimal sea view. The designers researched extensively on luxury yacht interiors and life under the sea. The design concept emerged as a bold experiment, mixing traditional and contemporary turquoise sensations with a unique feel, focusing on high impact drama and space.

To evoke a sense of serenity and boldness, like a yacht cruising atop dynamic waves, the overall space and form are shaped by dynamic wavy lines starting from the front 12m cantilevered box where 'Deep Bar' is located on the 4th floor accessible by a glass lift. The invisible waves continually run through lobby space where the lines are made tangible by the sheer size of the 'Wave Wall'. These undulating trajectories are continually utilised, as a tool to shape specific design elements that weave their way through the curved periphery of the pool, restaurants, bars, guestrooms with fascinating sexy open bathrooms, and d'spa.

Rounded beds float in the centre of each room on island-like rugs, while in some suites the luxury cabin motif is reinforced architecturally by placing the bed on a rounded balcony, overlooking the lounge. Strong

dusit**D2**
baraquda
pattaya

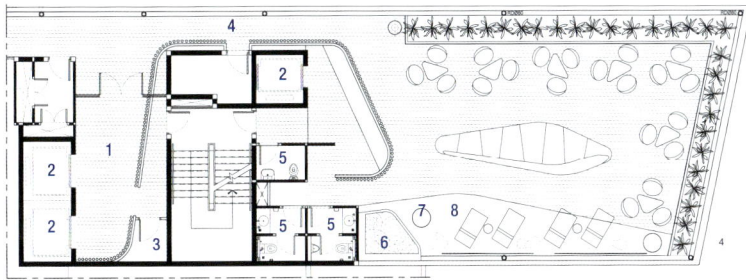

1. Entry   1. 入口
2. Lift   2. 电梯
3. Store   3. 储物间
4. Timber deck   4. 竹制甲板
5. W.C.   5. 卫生间
6. Waterfall   6. 瀑布
7. Sculpture   7. 雕塑
8. Sandscape   8. 沙地景观

1. Hotel entrance
   酒店入口
2. Hotel pool area
   酒店泳池区
3. Interior detail
   室内细节
4. Sunset lounge plan
   日落酒吧平面图

oceanic theme can also be found in the bar, which shimmers in aquatic shades under watery light where an undulating ceiling adds drama to the experience. The blend in the design creates an appealing shape and texture that accentuate the themes of water, air and light.

The 300sqm luxury d'spa is the crowning experience of any stay at this intriguing destination. Inspired by the marine leisure lifestyle and in keeping with the overall aquatic theme of the hotel, d'spa is a resplendent, calming environment in pure white and bright blue. Controlled temperature, lighting and sound mean that the spa and wellness journey can be tailored to the comfort of each guest's requirements. Natural elements and ambient candles add to the overall atmosphere of serenity and peace. Treatment rooms are spacious and massage beds offer the comfort to risk falling straight to sleep, while the therapists work their magic.

坐落于海滨城市芭提雅充满生机与活力的地方，芭提雅都喜D2芭拉古达酒店为游客提供东南亚与众不同的度假体验。该酒店的特色有奢华客房，动态的会议空间，提供美味佳肴的餐厅，酷深酒吧，可以观赏日落美景的屋顶休闲吧，独特的水疗区，景色壮丽的池边小道，和其他一系列服务设施。

业主最初的理念是想营造一个充满冒险而又放松的氛围，象征着银色金属光泽的梭鱼。这个设计对于狭长而海景很少的地点来说是一个很大的挑战。设计师广泛地研究了奢华游艇的室内设计和海底生物。设计理念源自于一个大胆的尝试，将传统与当代绿松石的感觉相结合，集中于营造高冲击的戏剧与空间独特感。

为了唤起一种宁静而大胆的感觉，就像一艘游艇航行在海浪之中，整个空间与形态都由动态的波浪线条构成，波浪始于乘玻璃升降梯可达的四层酷深酒吧前12米长的悬臂式盒子。无形的波浪继续延伸到大堂，又由陡峭的波浪墙形成有形的。这种波浪起伏的轨迹线继续延续着，作为一种形成特殊设计元素的手段，将蜿蜒的泳池连，餐厅与酒吧，带有令人迷醉的开放浴室与d'spa的客房交织在一起。

圆形床放置在每间客房中央，前面是一块岛型的地毯，营造了一种漂浮在海上的感觉，在有些套房里，将床放置在圆形的阳台上，可以俯瞰休闲吧，更加衬托了这种奢华客舱的主题。这种强烈的海洋主题也可见于酒吧之中，水形的光线下闪烁着水生物形的微光，波浪起伏的天花板更增加了戏剧性的体验。这种混合设计创造的形状和质地极具感染力，更加增强了设计的主题：水流、空气和光线。

300平方米的奢华d'spa是这个迷人之地上的最高体验。受海生物闲适的生活方式启发，并保持与整个酒店海洋主题的一致性，d'spa以纯白与湖蓝色营造出一种优雅而宁静的氛围。可控制的温度，灯光与音效意味着这里的水疗与健康之旅会满足每位客人的舒适需求。天然元素与周围的烛台更增添了宁静与平和之感。宽敞的理疗室里有舒适的按摩床，客人在理疗师的魔力下很可能会安然入睡。

5. Pool and deck
泳池与甲板区

6. Hotel façade
酒店外观

7. Sunset lounge
日落酒吧

8、9. Hotel lobby
酒店大堂

10. Wave wall
防浪墙

11. Level 1 floor plan
一层平面图

12-14. Deep Bar
深海主题酒吧

15. Level 4 floor plan
四层平面图

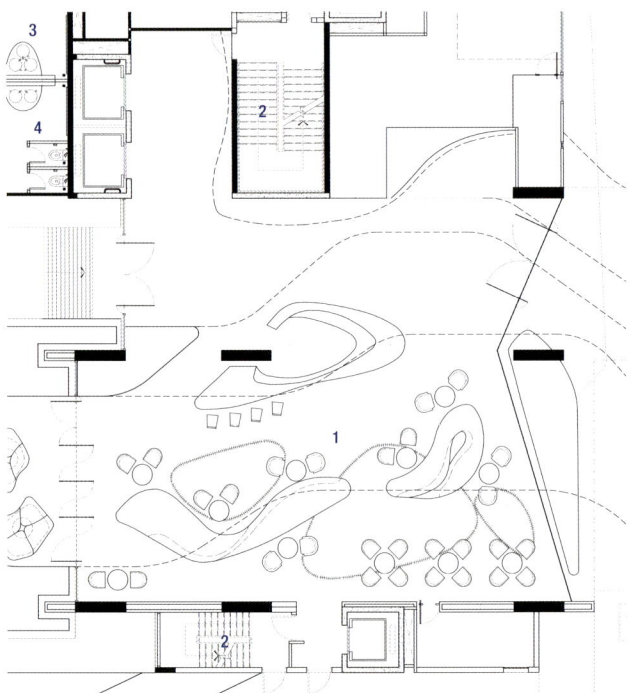

1. Lobby
2. Staircases
3. Male toilet
4. Female toilet

1. 大厅
2. 楼梯
3. 男卫生间
4. 女卫生间

1. Bar
2. Banquet seating
3. Bar tables & stools
4. Drink rail
5. Storage

1. 酒吧
2. 宴会厅就餐区
3. 酒吧就餐区
4. 饮品区
5. 储存区

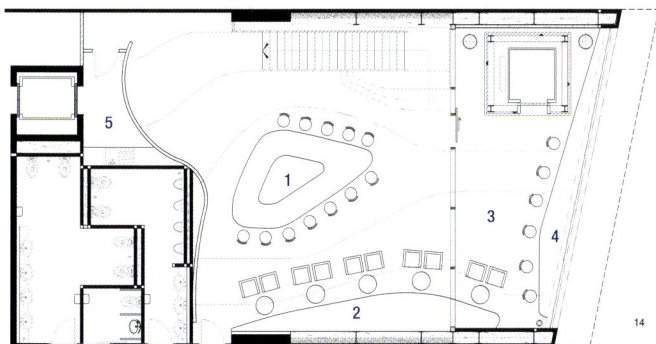

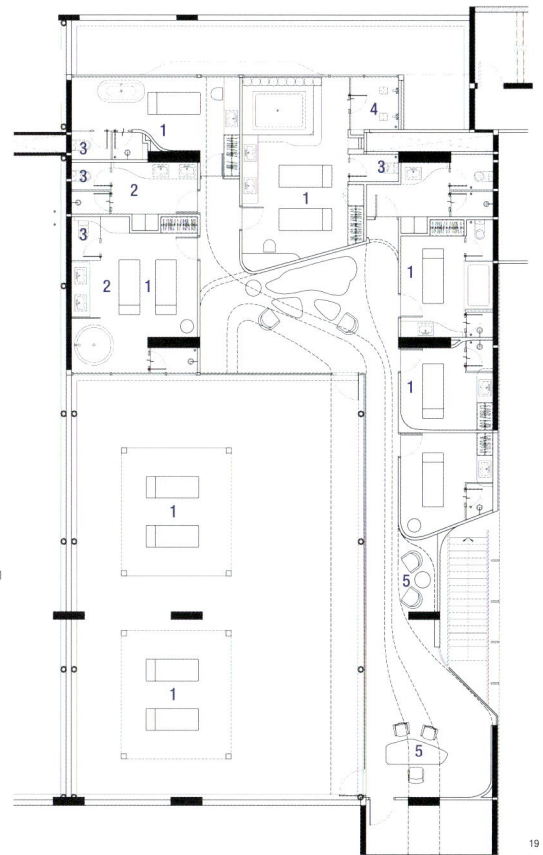

16. Pool view
    泳池景观

17. s.e.a restaurant
    s.e.a 餐厅

18. d'spa reception
    d'spa接待处

19. d'spa plan
    d'spa平面图

1. Treatment room
2. Bathroom
3. Toilet
4. Shower
5. Rest area

1. 治疗室
2. 浴室
3. 卫生间
4. 淋浴间
5. 休息区

20. d'spa exterior
    d' spa室外区

21~23. d' spa treatment room
       d' spa治疗室

24. Living room
    客厅

25. Presidential suite plan
    总统套房平面图

26. Guestroom
    客房

27. Living room
    客厅

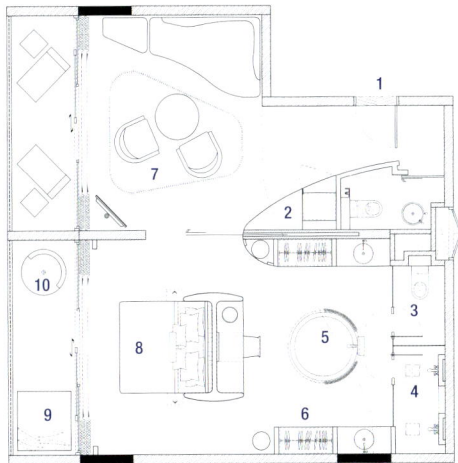

1. Entry
2. Bar
3. Toilet
4. Shower
5. Bathtub
6. Wardrobe
7. Living room
8. Bedroom
9. Outdoor sofa bed
10. Suspended seat

1. 入口
2. 酒吧
3. 卫生间
4. 淋浴间
5. 浴缸
6. 衣柜
7. 起居室
8. 卧室
9. 室外沙发床
10. 悬浮座椅

Phulay Bay, a Ritz-Carlton Reserve

普蕾湾丽嘉酒店度假村

Completion date: 2009
Location: Krabi, Thailand
Designer: P Landscape
Photographer: Christopher Cypert
Area: 4,500sqm

竣工时间：2009年
项目地点：泰国，喀比
设计师：P景观设计事务所
摄影师：克里斯多夫·赛博尔特
项目面积：4,500平方米

This is a resort that learns through the grace of Thai heritage by highlighting the art, architecture, customs and beliefs of Thai culture by exploring Phulay Bay. It ventures to a captivating corner of the world, where glistening sands and blue skies converge with the Andaman Sea, and where the charm and natural beauty of Thailand blends with an aura of serenity and discovery. A picturesque retreat, Phulay Bay is situated in Krabi, one of Thailand's emerging holiday destinations located on the Andaman Sea. Phulay Bay has a peaceful realm of contemporary Thai décor and gracious service in villas offering private tropical garden or ocean views. Set on the

shores of the Andaman Sea, Phulay Bay offers unspoiled views of the naturally rugged shoreline and dramatic limestone karsts, which make up this beautiful part of southern Thailand.

Each restaurant is served with flawless Thai hospitality in settings that reflect the unique culture and architecture of Thailand. Exquisitely appointed restaurants are adjacent to a serene lotus pond overlooking the Andaman Sea and its exotic islands.

Phulay Bay accommodations are luxuriously designed with a modern interpretation of contemporary Thai style and wonderfully complemented by traditional Thai hospitality. Spacious walk-in wardrobes prepared with comfortable cotton and Thai silk robes, as well as beach accessories and other necessities, are featured in every villa, as are oversized beds with luxurious pillows and exquisitely fine linens.

Phulay Bay is of great renown as the first ESPA resort destination Spa in Thailand – a unique oasis offers one-of-a-kind design and experience. Blending the soothing rhythms of the Andaman Sea with extraordinary Spa treatments, ESPA at Phulay Bay brings us into a blissful state of relaxation. It offers a unique selection of Spa therapies that enliven the senses and revitalise the spirit.

普蕾湾度假村汲取了泰式传统的优雅，凸显艺术、建筑、风俗和泰国文化理念。在这个世界上迷人的角落，闪耀的沙子和蔚蓝的天空与安达曼海汇聚在一起，泰国的魅力和自然美景与宁静、探索的氛围相互结合。

风景如画的普蕾湾度假村位于泰国安达曼海边新兴的度假胜地——喀比。普蕾湾是融合现代泰式装饰和高品质别墅服务的安宁国度。别墅配有私人热带花园或享有海洋美景。普蕾湾坐落在安达曼海边，为游客提供了自然崎岖的海岸线和生动的喀斯特地貌，体现了泰国南部的特色景观。

每间餐厅都以完美无瑕的泰式待客之道来接待宾客，反映了泰国独特的文化和建筑特色。精致的酒店紧邻宁静的荷花池，俯瞰着安达曼海和独特的岛屿。

普蕾湾度假村结合了现代泰式风格和传统泰国酒店特色，尽显奢华。每座别墅宽敞的步入式衣柜内配有舒适的棉质和泰式丝质长袍、沙滩用品和其他必需品，同时，别墅内还配有超大的双人床、奢华的枕头和精致的床单。

度假村拥有泰国首家ESPA温泉水疗中心——提供独一无二设计和体验的绿洲。ESPA混合了安达曼海舒缓的节奏和非凡的温泉理疗，让人们沉浸在幸福的轻松感之中。各种独特的温泉疗法能够唤醒人们的感官，令人重获新生。

1. Pool and relax pavilion
   泳池与休闲亭
2. Infinity pool
   无边泳池
3. Pool and relax pavilion
   泳池与休闲亭
4、5. Landscape around
      周围景观
6. Resort plan
   度假村平面图
7. Terrace lounge
   露台吧
8. Relax pavilion
   休闲亭

1. Arrival gate
2. Tarn court
3. Arrival pavilion, fitness centre, retail shop, function room
4. Reception pavilion
5. Spa
6. Club house
7. Pool
8. Pool restaurant and ball
9. Pavilion
10. Sunset lounge bar
11. International restaurant
12. Thai restaurant

1. 抵达区大门
2. 冰斗湖球场
3. 到达亭，健身中心，零售商店，功能间
4. 接待亭
5. 水疗中心
6. 俱乐部
7. 游泳池
8. 泳池餐厅与宴会厅
9. 凉亭
10. 落日休闲酒吧
11. 国际餐厅
12. 泰式餐厅

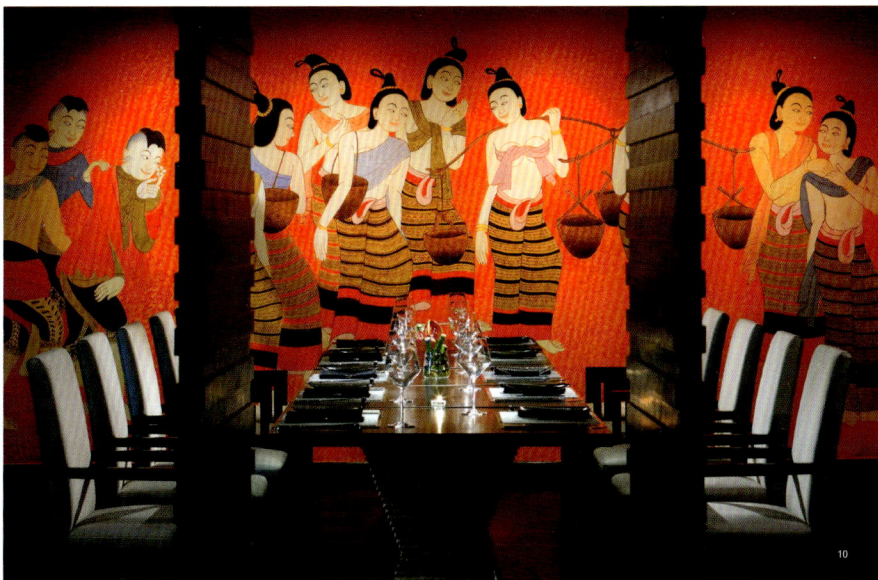

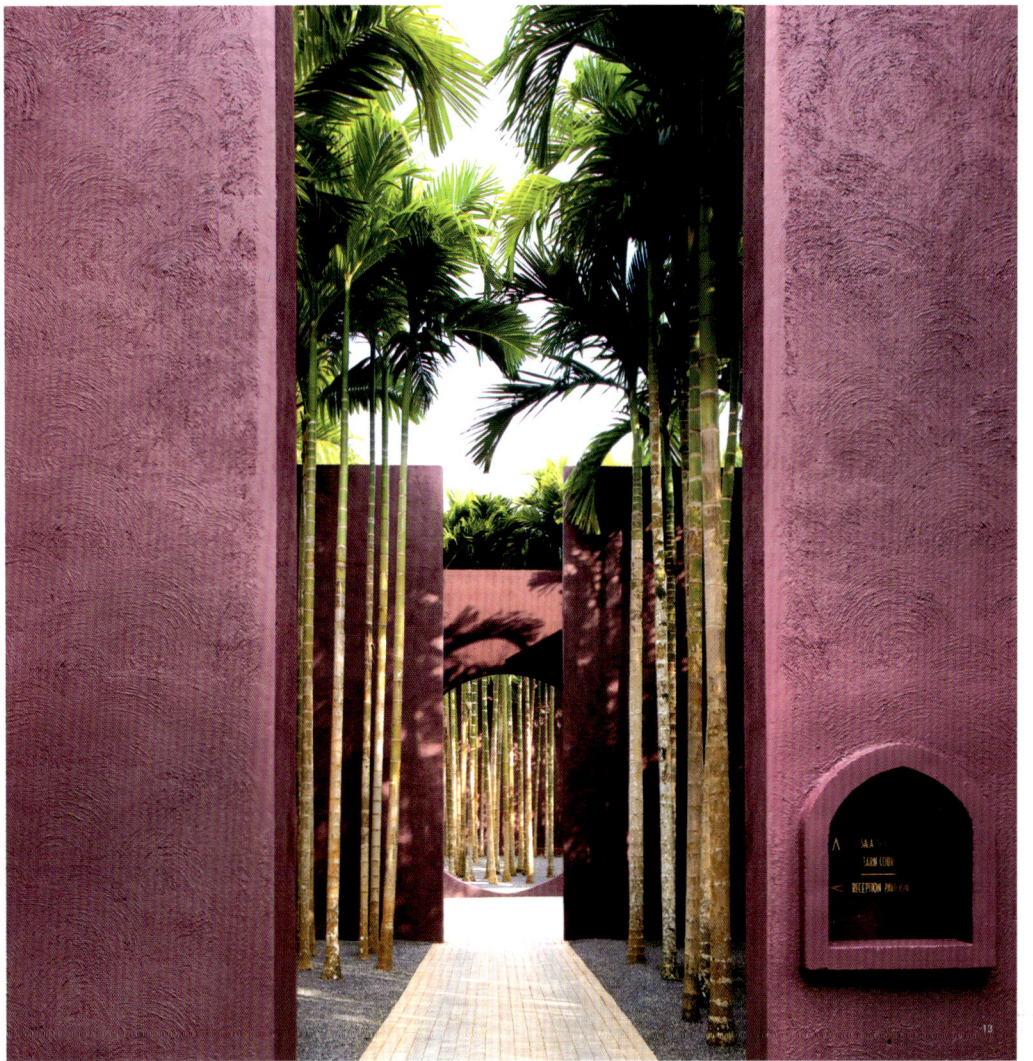

12. Entrance
    入口
13. Passageway
    通道
14. Spa pool
    水疗池

15. Lobby lounge
    大堂吧
16. Reception
    接待处
17. Spa relaxation area
    水疗中心放松区

18. Spa pool
    水疗池
19. Spa pool and relaxation area
    水疗池与放松区
20. Spa treatment room
    水疗中心治疗室
21. Living room
    客厅
22. Terrace
    露台
23. Guestroom
    客房

24. Guestroom
客房
25. Relaxation area
休闲区
26. Bathroom
浴室

Sofitel Krabi Phokeethra Golf & Spa Resort

索菲特喀比佛吉拉高尔夫水疗度假村

Completion date: 2006
Location: Krabi, Thailand
Designer: Choochart Polakit
Photographer: Jirasak Thongyuak

竣工时间：2006年
项目地点：泰国，喀比
设计师：丘查尔特·波拉基特
摄影师：吉拉萨克·东宇亚克

Sofitel Krabi Phokeethra Golf & Spa Resort is a majestic, luxury resort with 276 guest rooms, each with spacious balconies overlooking the ocean, towering limestone cliffs and lush and tropical gardens. This resort is designed and inspired by Choochart Polakit, presenting the magnificent concept of timeless colonial style with dark paneled wood, soaring arches and lofty high-vaulted ceilings. Rooms and suites are light and airy with polished teak floors, colonial-style wood furnishing, luxurious marble bathtubs and large private balconies overlooking the ocean, pool and garden.

Sofitel Krabi Phokeethra Golf & Spa Resort has been offering the ultimate in comfort and elegance by combining traditional colonial style architecture with the latest high-tech facilities in a tastefully-decorated and environmentally-friendly five-star resort. The hotel features 276 rooms and suites including 41 Superior Garden Rooms, 179 Superior Pool/Oceam Rooms, 33 Luxury Rooms, 5 Junior Suites, 10 Prestige Suites, 7 Opera Suites Pool Acccess and 1 Presidential Suite, in the elegantly designed and decorated in timeless French colonial style. All rooms and suites have a spacious bathroom, imported French bath products, and fully stocked mini-bar.

So SPA with L'Occitane at Sofitel Krabi Phokeethra Golf & Spa Resort created a serene spa oasis in a place as renowned for its natural beauty as it is for the ancient mythical legend of Naga, where all who visit are granted abundant blessings and peacefulness. So SPA with L'Occitane ushers you into this private sanctuary to begin your personal spa journey where everything has been conceived to soothe and harmonise your senses and to nature a deep feeling of wellbeing.

There is a modern business centre with full secretarial

1. Lobby
2. Activities centre
3. Grand Ballroom & Meeting Rooms
4. Café Vienna
5. So SPA with L'Occitane
6. Explorer Bar
7. Ristorante Venezia Restaurant
8. Maya Restaurant
9. White Lotus Restaurant
10. Cooking Class
11. Health & Fitness Centre
12. Kid's Club
13. Tennis Court
14. Opera Suite with direct access to the pool
15. Prestige Suite
16. Kid's Pool
17. Swimming Pool
18. Maprao
19. Koh Poda Pool Bar & Restaurant
20. Sala Thai (Wedding Pavillon)
21. Klong Muang Beach
22. Hotel Speedboat
23. Security Booth
24. Phokeethra Golf Krabi

1. 大堂
2. 活动中心
3. 大宴会厅与会议室
4. 维也纳咖啡厅
5. 欧舒丹So SPA水疗中心
6. 探险者酒吧
7. 威尼斯餐厅
8. 玛雅餐厅
9. 白莲花餐厅
10. 烹饪学习间
11. 保健与健身中心
12. 儿童俱乐部
13. 网球场
14. 直通泳池的歌剧套房
15. 威信套房
16. 儿童泳池
17. 游泳池
18. 马普罗餐厅
19. 康普德泳池酒吧与餐厅
20. 泰式大厅（婚礼亭）
21. 康芒海滩
22. 酒店快艇
23. 安全控制室
24. 喀比佛吉拉高尔夫球场

1. Resort plan
   度假村平面图
2. Swimming pool at night
   泳池夜景
3. Swimming pool at daytime
   游泳日间景色
4. Phokeethra Golf Krabi
   喀比佛吉拉高尔夫球场
5. Room terrace from balcony
   客房露台
6. Swimming pool and Sala Pavilion
   游泳池与婚礼亭

7. Phokeethra Grand Ballroom
   佛吉拉大宴会厅
8. Venezia Italian Restaurant
   威尼斯意式餐厅
9. White Lotus Thai — Indian Restaurant
   白莲花泰式餐厅
10. So SPA with L'Occitane
    欧舒丹 So SPA
11. Treatment room
    治疗室

services and four meeting rooms, ranging in size from 40 to 365 square metres. The pre-function room (140 square metres) seats up to 150 people; Phokeethra Meeting Room 1 (80 square metres) seats up to 70 people; Phokeethra Meeting Room 2 (40 square metres) seats up to 30 people; and The Grand Ballroom offers 365 square metres of space to seat up to 350 people for formal or informal receptions.

12. Hemingway Club
   海明威俱乐部
13. Opera Suite
   娥佩兰套房
14. Superior Room with king bed
   配有特大号床的高级客房

索菲特喀比佛吉拉高尔夫水疗度假村宏伟奢华，276套客房配有宽敞的阳台，俯瞰着海洋、峭壁和茂密的热带花园。度假村由丘查尔特·波拉基特负责设计，呈现了宏伟而经典的殖民地风格，深色木镶板、高耸的拱门和典雅的圆顶天花板尽显奢华。客房和套房配有光洁的柚木地板、殖民地风格木制家具、奢华的大理石浴缸和宽大的私人阳台，显得轻快而明亮。

索菲特喀比佛吉拉高尔夫水疗度假村提供顶级的舒适和高雅体验，结合了传统殖民地风格建筑和最新的高科技设施，打造出高品位而环保的五星级度假村。酒店的276套客房和套房包括41套高级花园客房、179套高级泳池/海景客房、33套奢华客房、5套普通套房、10套威信套房、7套歌剧泳池套房和1套总统套房。各个客房和套房都采用了高雅的设计和经典的法式殖民地风格。宽敞的浴室内采用法式洗浴产品，配有永远满载的小冰箱。

索菲特喀比佛吉拉高尔夫水疗度假村里的欧舒丹So SPA水疗中心打造了一片宁静的水疗绿洲，以古老而神秘的那迦文化为蓝本，为宾客提供满满的祝福和平和感。欧舒丹So SPA水疗中心引领客人进入私密的圣地，开始私人水疗之旅。全部设施都舒缓并抚慰着人们的感官，为人们带来深层次的幸福感。

度假村的现代商务中心配有全套的商务服务，四间会议室的规模从40平方米到365平方米不一。准备室（140平方米）可容纳140人；一号佛吉拉会议室（80平方米）可容纳70人；二号佛吉拉会议室（40平方米）可容纳30人；大型宴会厅则提供了365平方米的空间，可容纳350人，进行正式和非正式的接待。

SALA Phuket Resort and Spa

普吉岛莎拉水疗度假村

Completion date: 2009
Location: Phuket, Thailand
Designer: Department of ARCHITECTURE Co.Ltd.
Photographer: Mr.Wison Tungthunya
Area: 1,500sqm

竣工时间：2009年
项目地点：泰国，普吉岛
设计师：Department of ARCHITECTURE建筑设计有限公司
摄影师：维森·汤塞亚
项目面积：1500平方米

SALA Phuket Resort and Spa is a beach resort featuring villas with private swimming pools in 63 out of 79 units. Facilities includes beachfront swimming pools, a beachfront restaurant and bar, spa, gym, gift shop, and executive meeting facility. The design approach focuses on the sensational dimension of the experience that is embodied within the newly conceived space. The physical form of the architecture is modest rather than aggressive. The visual gives way to the experiential.

From entering the property at one end to arriving at the beachfront on the other end, space is conceived

as a continuous perceptual unfolding of experiences. This sequence of experience starts at the drop-off area where interior of the resort is partially revealed through a wooden screen. Behind the screen situates a reception area where an open landscape of the property is fully revealed. The final section begins at the beachfront public area where the pools and restaurants accompany an openness of the sea.

Within each villa, space is conceived as a privately enclosed living quarter. Spaces within the precinct flow freely between the interior and the exterior. An open plan seamlessly fuses the space of the bedroom, the wash area, the living space, the pool and the landscape into a continuous living experience.

The main public area is composed of two beach front swimming pools and a restaurant and bar. The two main pools are laid along the length of the beachfront and become a peaceful foreground of the entire public area.

Taking advantage of the tropical climatic condition, the restaurant is designed to operate as an open-air pavilion. The volume of the restaurant and bar has been programmatically stretched horizontally along the shoreline so as to place all the visitors along the

1. Guest parking
2. Staff parking
3. Back of house
4. Boardroom
5. Lobby
6. Villa
7. Front office
8. Spa
9. Shop
10. Kitchen
11. Restaurant & Bar
12. Lap pool
13. Fun pool

1. 客用停车场
2. 员工停车场
3. 后台区
4. 会议室
5. 大厅
6. 别墅
7. 前台
8. 水疗中心
9. 商店
10. 厨房
11. 餐厅与酒吧
12. 小型健身游泳池
13. 娱乐泳池

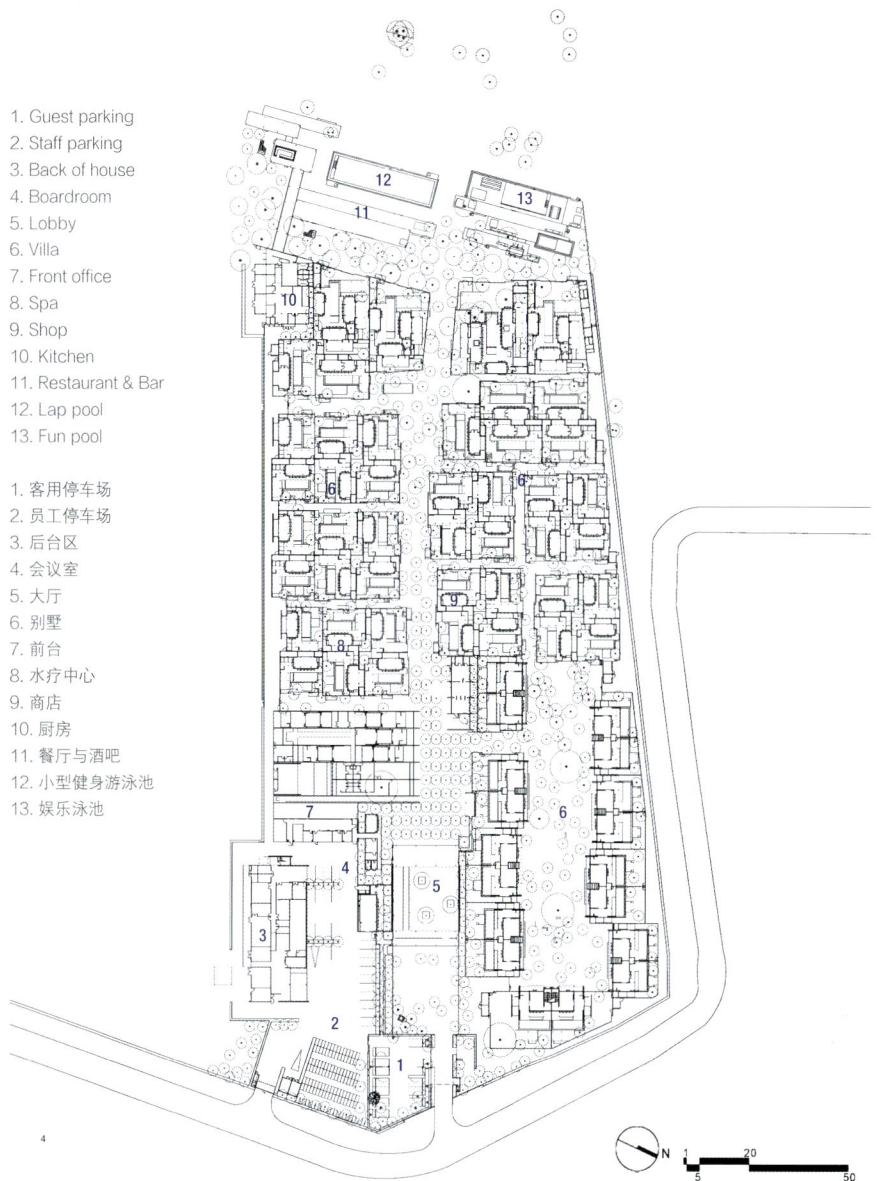

best view. Contrary to the grid system which usually demarcates a clear boundary line, structural columns are re-positioned at an unconstrained rhythm, thus creating an ambiguous edge—a sense of borderlessness. Consequently, the inside space and its surrounding environment is seamlessly merged into one.

Besides dining area on the ground, a new landscape of dining experience can be found on the restaurant's rooftop. Dining area is depressed into a reflecting pond forming sunken dining pods. By replacing conventional guard rails with water boundary, visual obstruction is removed. The space is extended outward indefinitely.

For the complexity and scale of the project, within the general design framework that ties all the programmatic elements together, there are yet more layers of architectural qualities to be discovered once one gets closer to the different parts of the programme. Each part of the project has its own architectural particularity whether it is the beach front public space, the restaurant and bar, the lobby or the villa, all through smaller components as a restroom.

普吉岛莎拉水疗度假村是一家海滨度假村，建有79间别墅，其中63间是特色的私人泳池别墅。其他设施有海滨游泳池、海滨餐厅与酒吧、水疗中心、健身中心、礼品店以及行政会议设施。设计手法着重于新构思空间中体现的经验的广大维度。建筑的整体架构是温和而不激进的，更加强调经验而不是视觉效果。

度假村的一端为入口，另一端通往海滩，空间的构思是一种连续的感性经验的展开。这一系列的经验开始于即停即离区，通过一个木制屏风，度假村的室内开始慢慢展开。屏风的后面是接待处，这里充分展现了度假村的开放景观。而这一系列的经验结束于海滨的公共区，这里有海滨泳池与海滨餐厅。

每幢别墅内的空间都被构思成一个私人独立生活区。生活区内的空间体现了室内外的自然流动性。开放式的布置将卧室、盥洗室、起居室、泳池与景观各个空间天衣无缝地融合在一起，形成一个连续性的生活经验。

主公共区由两个海滨泳池和一个海滨餐厅和酒吧构成。这两个主泳池是沿海设置的，位于整个公共区域的宁静而显著的地方。

充分利用热带的气候条件，将餐厅设计成露天的凉亭。规模庞大的餐厅与酒吧以各种主题形式沿海岸线水平延伸，是为了让游客观赏到最美的景色。不同于坐标往往要划清界线，框架结构柱以一种无拘无束的节奏放置着，创造了一个模糊边缘，给人一种无边无界之感。

除了一层餐饮区之外，餐厅的顶层还有一个新的景观餐饮体验区。餐饮区置于倒影池中，形成了一个个凹型餐饮区。由水界替代了传统的防护轨，这样可以扫清一切视觉障碍，使空间无限延伸。

考虑到项目的复杂性与规模，在总体设计架构中，将所有标题性元素结合在一起，然而一旦更加接近不同部分，还可以发现更多层次的架构品质。该项目每个部分的架构都有其独特性，无论是海滨公共区、海滨餐厅与酒吧，还是大堂、别墅，以至到如盥洗室这样更小的部分。

1、2. Lobby
　　大堂
3. Restaurant
　　餐厅
4. Resort plan
　　度假村平面图
5~8. Restaurant
　　餐厅

9. Spa entrance
　　水疗中心入口
10、11. Spa corridor
　　水疗中心通道
12、13. Spa reception
　　水疗中心接待处

14

14. Pool villa
泳池别墅

15. Duplex villa pool area
双层别墅泳池

16. Duplex villa bedroom
双层别墅客房

17. Two-beroom pool villa suite
两居室泳池别墅套房

18. Villa pool and deck
别墅泳池与甲板区

19. Villa façade
别墅外观

20. Villa pool
别墅泳池

21. Pool villa bedroom
泳池别墅客房

22. Pool and relax area
泳池与休闲区

23. Pool and deck
泳池与甲板区

15

16

Mövenpick Resort & Spa Karon Beach Phuket

普吉岛慕温匹克水疗度假村

Completion date: 2006
Location: Phuket, Thailand
Designer: P49 Design and Associates (Interior Design)
Bill Bensley of Bensley Design Studio Bangkok
(Hotel Landscape design)
Photographer: Ken Seet, Hamilton Lund
Area: 81,600sqm

竣工时间：2006年
项目地点：泰国，普吉岛
设计师：P49联合设计事务所（室内设计）
班斯利泰国设计工作室的比尔·班斯利（景观设计）
摄影师：肯·希特，汉密尔顿·伦德
项目面积：81,600平方米

The Mövenpick Resort & Spa Karon Beach Phuket is located on the west coast of Phuket, one of Thailand's most popular destinations. It is bordered by Karon Beach, well-known for the beautiful fine white sand and relaxed atmosphere. The resort harmoniously blends Thailand's gentle way of life with its Swiss tradition of hospitality and attention to detail to create a unique holiday product for discerning travellers.

Staying at the resort is like living in a tropical garden. It has one of the most beautiful gardens of any hotel in Phuket, designed by the renowned landscape designer Bill Bensley, who effortlessly combined

nature with culture to create a truly unique local experience. He shaped the gardens around existing trees and complemented them with water features and exotic plans. It creates an amazing experience of scents and colours. Spacious and elegant villas and four swimming pools blend into the lush landscape. Traditional Thai salas have been set up throughout the gardens, inviting hotel guests to enjoy the serenity of their surroundings.

Giant Thai-Chinese style birdcages, sculptures of the mythical Thai Kinnaree and wrought iron lanterns reminiscent of the tin mining past, give the resort a distinctively whimsical feel. The atmosphere is perfect for any sort of guests: business travellers, honeymoon couples, or families. The broad range of room types, from Deluxe Garden View Room to Two-Bedroom Garden Suites and Beachfront Two-Bedroom Residences, caters to the needs of all guests.

The resort has 175 elegant rooms, all decorated in traditional Thai style, with beautiful views of the resort grounds or the sea. Rooms are equipped with all the amenities expected at a five-star resort. It also offers 159 villas nestled amidst the resort's garden. Traditional materials including wood, thatched roofs

1. Resort bird's eye view
   度假村鸟瞰图
2. Resort main pool
   度假村主泳池
3. Villas & Coconut Grove Pool
   别墅与椰林泳池
4. Hotel lobby
   酒店大堂

5. Spa entrance
   水疗中心入口
6. Spa fountain
   水疗中心喷泉
7. Resort plan
   度假村平面图
8. Spa Dip pool and outdoor massage sala
   水疗中心水疗池与室外按摩亭

9. Spa treatment room
   水疗中心治疗室
10. Plunge Pool Villa
    跌水潭别墅

and an open design create a harmonious blend between the manicured gardens and luxurious interiors. The resort features Deluxe Garden View Rooms, Deluxe Ocean View Rooms, Garden Villa, Deluxe Plunge Pool Rooms, Deluxe Garden Villas, Plunge Pool Villas, One-Bedroom Garden Suites, Penthouse Plunge Pool Villas, Family Villas, Two-Bedroom Family Suites and Beachfront Two-Bedroom Butler-Serviced Residences. The latest additional to the resort is Beachfront Two-Bedroom Butler-Serviced Residences which are ultimate in luxury. Five elegant contemporary buildings with 30 oceanfront units offer superb views of Karon Beach and the Andaman Sea.

1. Lobby
2. Pool bar
3. Grand ballroom
4. Meeting room
5. Boardroom
6. Pacifica
7. Kinnaree
8. Kiosk
9. Gift shop
10. Transportation desk
11. Tour desk
12. Bell counter
13. Play zone
14. Fitness centre
15. Café studio
16. Sand Bar
17. OrientAsia
18. El Gaucho
19. ATM machine
20. Spa
21. Garden Rooms and Suites
22. M venpick Villas
23. Residences

1. 大堂
2. 泳池酒吧
3. 大宴会厅
4. 会议室
5. 董事会议室
6. 帕西菲卡餐厅
7. 肯纳里餐厅
8. 凉亭
9. 礼品店
10. 交通服务台
11. 旅游服务台
12. 礼宾部
13. 游乐天地
14. 健身中心
15. 咖啡室
16. 沙滩酒吧
17. 东亚餐厅
18. 高卓餐厅
19. 自动提款机
20. 水疗中心
21. 花园客房与套房
22. 幕温匹克别墅
23. 住宅区

11. Pacifica Restaurant
    帕西菲卡餐厅
12. Two-bedroom Family Suite
    两居室家庭套房
13. One-Bedroom Garden Suite bedroom
    一居室花园套房客房
14. Deluxe Garden Villa bedroom
    豪华花园别墅客房
15. Penthouse Plunge Pool Villa bedroom
    跌水潭别墅客房
16. Garden Villa bedroom
    花园别墅客房
17. Deluxe Garden View bedroom
    豪华花园景观客房
18. Family Villa bedroom
    家庭别墅客房

普吉岛慕温匹克水疗度假村坐落于泰国最受欢迎的普吉岛西海岸，比邻卡伦海滩，以迷人的精细白沙和休闲放松的氛围文明。该度假村将泰国温和的生活方式与瑞士注重细节的酒店传统和谐地融合在一起，为独具慧眼的游客创造了一个独一无二的度假胜地。

置身于该度假村犹如生活在一个热带花园里。它建有普吉岛酒店中最美的花园，由著名的景观设计师比尔·班斯利设计，他将自然与文化轻松结合，创造了一个当地的独特体验。花园周围种满了树，并以水文要素与异国情调点缀其中，创造了一个味觉与视觉上的惊奇体验。宽敞典雅的别墅和四个游泳池也被融入进这苍翠繁茂的景观之中。传统泰式小亭笼遍布花园之中，吸引游客在此享受周围的宁静。

中泰风格的大鸟笼，神秘的泰国肯纳里雕塑，和锻铁灯笼会令人追忆过去的锡矿时代，给度假村增添了一种怪诞色彩。这里的绝佳的氛围适合各类游客：商务旅行者、

蜜月旅行的夫妻、还有家庭旅行者。客房也有各种类型，包括豪华花园景观客房，两居室花园套房，还有两居室海景住宅，适合各种需求的客人。

度假村共有175间客房，装饰典雅，具有传统泰式风格，均可观赏海景或度假村景观。房间配置有5星级度假村所具有的各种服务设施。度假村有159幢别墅，分别隐建于度假村的花园中。传统材料如木材、茅草屋顶等的使用，和开放的设计，使修剪整齐的花园与奢华的室内和谐统一。度假村的特色客房有：豪华花园景观客房、豪华海景客房、花园别墅、豪华跌水潭客房、豪华花园别墅、跌水潭别墅、单卧室花园套房、阁楼跌水潭别墅、家庭别墅、两居室家庭套房和两居室海景住宅。两居室海景住宅是度假村最新增加的客房，装饰极其奢华。5幢典雅的现代风格建筑，设有30间海景房，是观赏卡伦海滩和安达曼海景色的绝佳之地。

Zazen Boutique Resort & Spa Resort

渣任精品水疗度假村

Completion date: 2009 (renovation)
Location: Koh Samui, Thailand
Designer: Jeri De Jongh, Thitima Fathaveeporn,
Alexander Andries
Photographer: Claudio Cerquetti
Area: 7, 000sqm

竣工时间：2009年（翻新）
项目地点：泰国，苏梅岛
设计师：杰里·德容，西提玛·发瑟维波恩，亚历
山大·安德里亚斯
摄影师：克劳迪奥·契尔凯蒂
项目面积：7,000平方米

Zazen Boutique Resort & Spa stands out as one of the first resorts on the island, and more importantly one of the first true 'Boutique' resorts in Thailand. Through these long years, the resort has been built, renovated, redesigned and reinvented, ending with a final result that can only be described as romantic, intimate and natural.

The beautiful bungalows and facilities are designed with taste and made of natural materials, mainly stone and wood, with colours that recall the earth matching the warm welcoming feeling. The style, mainly Thai and Balinese, perfectly blends with

the Moroccan lanterns spread out above the whole resort. The Spa welcomes guests in a more Oriental-Indian architecture, which matches with some of the Ayurvedic therapies proposed among the treatments, and immediately provides visitors with a deep sense of relaxation and well being.

Zazen Restaurant is definitely today's most romantic, glamorous and beautiful dining beachfront location on the island. High ceilings with elegant draperies, red Moroccan lanterns, hundreds of candles everywhere and soft jazz music playing in the background blended with the sound of the waves, provide the perfect atmosphere for a romantic dining experience.

When entering the restaurant, the first thing that will attract your attention is the underground wine cellar, which has glass ceiling that allows anyone to look into it from above and understand that wine is something taken seriously here. Bottles rest on beautiful wine racks in brick alcoves at near-perfect storing conditions, and this is where you can find some great wines and vintages.

When working on the new restaurant's project, Alex and Thitima thought it would be useful to create a new venue dedicated to private events such as

1. Reception & Courtyard
2. Beachfront Deluxe Bungalows
3. Garden Deluxe Bungalows
4. Garden Villas
5. Le Spa Zen
6. Zazen's Boutique & Business Lounge
7. Zazen Restaurant
8. Le Rouge Lounge & Bar
9. Swimming pool
10. Recreation Sala

1. 接待处与庭院
2. 海滨奢华廊房
3. 花园奢华廊房
4. 花园别墅
5. 禅式水疗中心
6. 渣任精品店与商务休闲吧
7. 渣任餐厅
8. 鲁日休闲吧与酒吧
9. 游泳池
10. 娱乐大厅

weddings, cocktail parties and private dining. Assisted by Mr. Jeri De Jongh, they gave birth to Le Salon de Ti. The venue gets its name from Khun Thitima's nickname, Ti, and from the word 'salon' which in French literally means 'living room'. This is Ti's living room, where she personally welcomes each guest visiting just like in her own house.

渣任精品水疗度假村是苏梅岛上首批度假村里最出众的，也是泰国首家真正意义上的精品度假村。这些年来，该度假村从建造、翻新，到重新设计，最后彻底改造成现在这个浪漫、亲密、而自然的度假之地。

美丽的度假小屋和各种设施的设计独具品位，并由天然材料建造而成，主要的材料是石头和木材，颜色与大地和谐统一，并能营造出热情好客的氛围。遍及整个度假村的摩洛哥花灯与度假村主要的泰国风格和巴黎岛风格完美融合。水疗中心的建筑是东方印度风格，这种风格与理疗法中的阿育吠陀疗法相匹配，并会立即给游客带来一种放松与幸福感。

渣任餐厅是如今岛上最浪漫、富有魅力而又迷人的滨海餐饮区。高高的天花板装饰典雅，红色的摩洛哥花灯，成百上千的蜡烛随处可见，轻爵士乐伴随着海浪的声音，构成绝佳氛围，为游客提供浪漫的就餐体验。

进入餐厅首先映入眼帘的是地下酒窖，玻璃天花板可以让人从上面直接看到里面，并且了解到这里的酒是得到了充分重视的。酒瓶放置在精美的酒架上和砖砌的壁龛格子里，在绝佳的条件下，储藏着美味的葡萄酒。

在新的餐厅项目设计中，亚历山大和西提玛都觉得有必要创造一个全新的专门提供婚礼、鸡尾酒会和私人聚餐的私密场所。在杰里·德容的协助下，他们这样做了，并建造了德西沙龙。 这个名字来源于坤恩·西提玛的绰号西，和法语中意为"起居室"的"沙龙"一词。这里是西的起居室，在这里她如同在家里一样迎接每一位客人。

1. Zazen's reception
   渣任度假村接待处
2. Zazen Restaurant's entrance
   渣任餐厅入口
3. Resort plan
   度假村平面图
4. Zazen's pool & restaurant
   渣任泳池与餐厅
5. Spa area
   水疗区
6. Le Salon de Ti
   缇沙龙
7. Zazen Restaurant's
   beachfront terrace
   渣任餐厅海滨露台
8. Le Salon de Ti
   缇沙龙
9. Zazen restaurant's wine cellar
   渣任餐厅酒窖

10. Zazen Restaurant
    渣任餐厅

11. Spa pendant lamp
    水疗区吊灯

12. Spa fountain
    水疗区喷水池

13. Spa relaxation area
    水疗休闲区

14. Le Salon de Ti tea lounge
    缇沙龙茶吧

15. Spa tea lounge
    水疗区茶吧

16. Spa relax area
    水疗休闲区
17. Spa corridor
    水疗走廊
18. Spa relax area entrance
    水疗休闲区入口
19. Spa treatment room
    水疗治疗室
20. Zazen's Beachfront Bungalows
    渣任海滨别墅
21. Beachfront Deluxe Bungalow bedroom
    海滨奢华别墅客房
22. Zazen's Garden Villa bedroom
    渣任花园别墅客房

Sheraton Maldives Full Moon Resort & Spa

马尔代夫满月岛喜来登度假酒店

Completion date: 2008
Location: Male, Maldives
Designer: P49 Design and Associates
Photographer: Sheraton Maldives Full
Moon Resort & Spa
Area: 200 sqm

竣工时间：2008年
项目地点：马尔代夫，马累
设计师：P49联合设计事务所
摄影师：马尔代夫满月岛喜来登度假酒店
项目面积：200平方米

Sheraton Maldives Full Moon Resort & Spa is situated on its own island - one of 250 inhabited isles among the over 1,100 comprising the Republic of Maldives - and is surrounded by tropical turquoise waters and its own lagoon.

44 Beach-Front Deluxe Rooms are spacious sanctuaries nestled between the palm trees. Offering the option of ground or first floor rooms, each block of four rooms features a thatched roof, and all rooms boast a large balcony or terrace with stylish, cushioned outdoor furniture. Beach-Front Deluxe rooms are located near the beach with a choice of garden and

resort views or relaxing sea views.

Spacious thatched-roof Beach-Front and Island Cottages feature a private balcony and either views of our tropical gardens and lagoon, or sweeping ocean vistas and direct beach access. Privacy and romance are assured in the cosy bungalows with magnificent ocean views. Each stand-alone Bungalow is its own welcome respite with a thatched roof, sun terrace, and direct stair access to the clear, turquoise waters below. Four Water Villas and one Water Suite offer a lavish tropical retreat. Treat yourself to the experience of an over-water bungalow, but with added space and seclusion, upgraded amenities, and beautiful rattan furnishings.

Combining a modern interior design, 20 private, chic Ocean Villas boast uninterrupted ocean views and luxurious outdoor areas with plunge pool, dining area, hammock, and more. The private deck features an individual plunge pool with a daybed and an outdoor rain shower. A dining area, a hammock, and garden seating area with a personal BBQ complete the ultimate luxury outdoor living experience. Inside, calming colours and Balau wood flooring create a distinctly Maldivian décor.

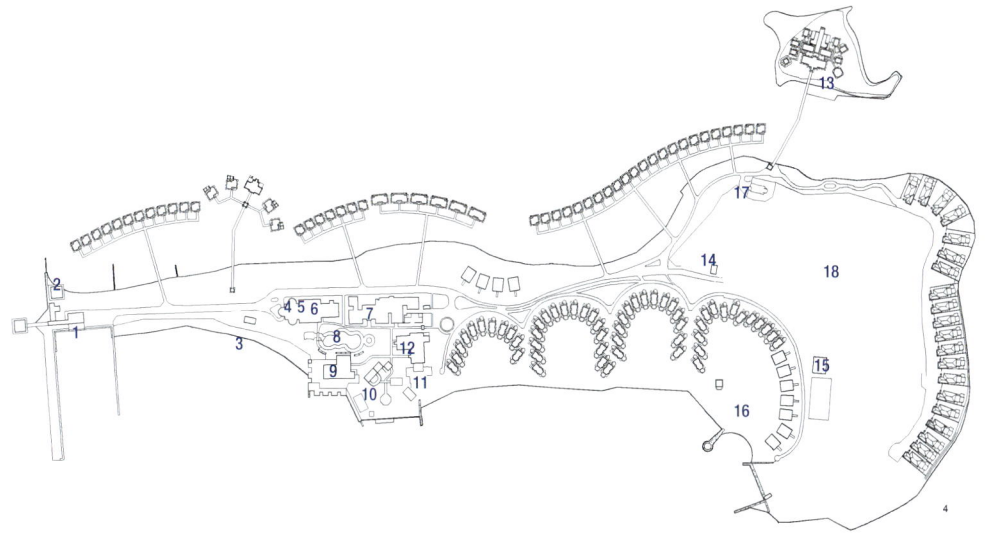

1. Welcome pavillion	10. Sea Salt	1. 迎宾亭	10. 海盐餐厅
2. Dive centre	11. Sand Coast	2. 潜水中心	11. 海滨沙滩餐厅
3. Water sport centre	12. Baan Thai	3. 水上运动中心	12. 博安泰式餐厅
4. Lobby	13. Shine spa for Sheraton	4. 大堂	13. 喜来登阳光水疗中心
5. Jewellery shop & boutique	14. Clinic	5. 珠宝店与精品店	14. 诊所
6. Drifters	15. Fitness centre & Tennis court	6. 漂流区	15. 健身中心与网球场
7. Feast	16. Beach Bar	7. 宴会厅	16. 海滨酒吧
8. Swimming pool	17. Sheraton Adventure Club	8. 游泳池	17. 喜来登冒险俱乐部
9. Anchorage Bar	18. Star village – for associates only	9. 安克雷奇酒吧	18. 明星村（只限会员）

The new Shine Spa is genuinely unique and truly sensational. It has taken inspiration from both the East and the West, and combined them into a special treatment experience. With treatments originating all the way from China, Thailand, India, Arabia to the Mediterranean, it will provide the guest a truly magical spa experience.

马尔代夫满月岛喜来登度假酒店坐落于自己的私人岛屿上——由1,100多座岛屿组成的马尔代夫共和国共有250座岛屿有人居住，这座美丽的小岛就是其中之一，四周环绕着碧蓝的热带海洋及迷人的环礁湖。

44间宽敞的海滨豪华客房掩映在葱郁繁茂的棕榈树间，堪称舒适安宁的避世天堂。客房位于酒店一楼或二楼，每四间客房组成一个区域，并采用茅草屋顶设计，所有客房均设有宽敞的阳台或露台，并配以带软垫的时尚室外家具。靠近海滩的海滨豪华客房可提供花园和度假酒店景观或令人心旷神怡的海洋景观。

茅屋顶海滨别墅和海岛别墅空间宽敞明亮，配有私人阳台，可饱览热带花园和礁湖美景，或一望无际的海洋胜景，并可直通海滩。舒适小屋可提供私密而浪漫的宜人氛围，并可饱览波澜壮阔的迷人海景。每栋独立小屋均拥有自己的独特风格，并采用别致的茅草屋顶设计，且配有阳光露台和通向碧蓝海水的直达楼梯。

四间水上别墅和一间水上套房是极尽奢华舒适的热带幽居。您可在宽敞的水上别墅尽享悠闲惬意，升级服务设施和精美藤制家具将为您营造出世外桃源般的静谧氛围。

20座私密别致的海景别墅设计现代时尚，拥有壮观迷人的海景，并提供带泳池、餐饮区和吊床的豪华室外休闲区。私人露台上设有一座独立泳池、睡椅和室外淋浴间。餐饮区、吊床以及带有私人烧烤的花园休闲区，共同营造出极致奢华的户外生活体验。每座别墅均是放松休闲、享受洋无尽美景的绝妙居所。别墅之内，清新宜人的色彩和实木地板则渲染出马尔代夫独具特色的装饰风格。

全新的 Shine 水疗中心独特而舒适，为您带来极致感官享受。它将东西方传统的精髓理念融汇贯通，并运用到特色的护理之中。源自于中国、泰国、印度、阿拉伯和地中海地区的各种特色护理将为客人打造真正神奇的水疗体验。

1. Water pavilion exterior
   水亭外观

2. Water Bungalow sunset
   日落时分水上屋

3. Welcome pavilion exterior
   欢迎亭外观

4. Resort plan
   度假村平面图

5. Resort fountain
   度假村喷泉

6. Anchorage Bar
   安克雷奇酒吧

7. Water Bungalow terrace
   水上屋露台

8. Beachside dining pavilion
   海滨餐饮亭

9. Spa relaxation pool
   放松水疗池

10. Feast restaurant
宴会餐厅
11. Drifters Restaurant
漂流者餐厅
12. Beachfront villa
海滨别墅
13. Spa lobby
水疗中心大堂
14. Spa treatment room
水疗中心治疗室

15. Ocean Villa at sunset
日落时分海洋别墅
16. Lobby
大堂
17. Cottage semi-open bathroom
小别墅半开放浴室
18. Ocean Villa bathroom
海洋别墅浴室
19、20. Ocean Villa bedroom
海洋别墅客房

Paradise Island Resort & Spa

天堂岛水疗度假村

Completion date: 2008 (renovation)
Location: North Malé Atoll, Maldives
Designer: Tekton Design Associate Pvt. Ltd.
Photographer: Villa Hotels & Resorts
Area: 187,131sqm

竣工时间：2008年（翻新）
项目地点：马尔代夫，马累北环礁
设计师：Tekton联合设计公司
摄影师：维拉酒店与度假村集团
项目面积：187,131平方米

Paradise Island Resort & Spa is set on North Malé Atoll of the Maldives. This private island surrounded by the gorgeous sands and enchanting submarine realms, measures 931 metres in length and 201 meters in width. Paradise Island Resort & Spa is an offer of luxury, richly endowed with vegetation blazing with tropical blooms and displaying exotic views of the sea and beaches.

The grand reception hall is a place to relax in style. It's also the place to find out what's happening. A network of paved pathways cut through the vegetation interconnecting the luxurious accommodations, a

variety of restaurants, sports areas and a multitude of facilities that serve to make your holiday all the more complete.

Paradise Island Resort & Spa offers a variety of accommodations in a natural paradise of unspoiled beauty. The resort features 220 Superior Beach Bungalows that completely encircle the island and 40 Water Villas, 16 Haven Villas, 4 Haven Suites and 2 Ocean Suites built on stilts protruding out of turquoise waters of the lagoon, offering the ultimate in comfort and leisure. Each pagoda-style Superior Beach Bungalow faces the Bounty-like beach and a few steps away from the cashmere-soft shores. The Villas and Suites are exquisitely furnished with contemporary décor and each suite features its own private pool, Jacuzzi and outdoor dining area.

The Araamu Spa tucked away in a secluded corner of the island features 17 treatment rooms for massages, body treatments and facials, of which 4 rooms are exclusively for Ayurvedic treatment and 1 room for hydrotherapy treatment. Each of the private treatment room features outdoor courtyard, shower and flower bath. Any treatment here will make you think you have just stepped out of a heavenly sanctuary.

1. Around the resort island
   度假村周围景色
2. Resort plan
   度假村平面图
3. The haven coffee shop
   天堂咖啡厅
4. Beach view
   海滩景色

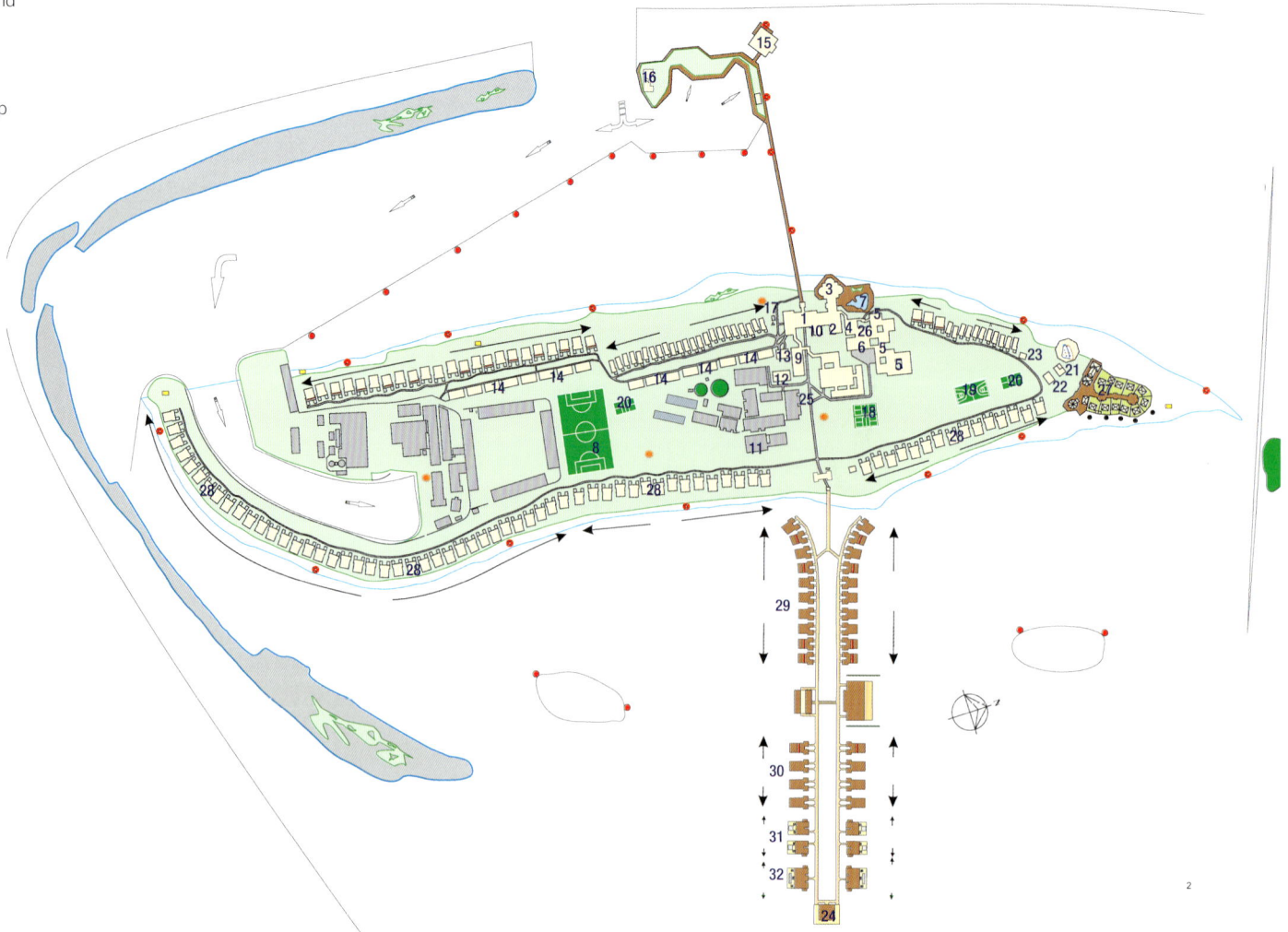

1. Reception/Office	1. 接待处/办公室
2. Conference Hall	2. 会议厅
3. Hulhangu Bar/Disco	3. 瑚瀚谷酒吧、迪斯科舞厅
4. Davani Coffee Shop	4. 达瓦尼咖啡店
5. Bageecha Restaurant	5. 芭吉餐厅
6. Bageecha Restaurant kitchen	6. 芭吉餐厅厨房
7. Swimming pool	7. 游泳池
8. Soccer field	8. 足球场
9. Sports centre/Photo shop	9. 运动中心、照相馆
10. Electronic shop/Jewellery shop & Souvenir shop	10. 电子商店、珠宝店和纪念品店
11. Squash and Badminton	11. 壁球场与羽毛球场
12. Fitness Club	12. 健身俱乐部
13. Clinic	13. 诊所
14. Garden Front Bungalow	14. 花园小屋
15. Sun Set Restaurant	15. 落日餐厅
16. Diving School	16. 潜水学校
17. Luggage Room	17. 行李间
18. Tennis Courts	18. 网球场
19. Basket Ball Court	19. 篮球场
20. Volley Ball Court	20. 排球场
21. Water Sports	21. 水上运动中心
22. Sails	22. 帆船中心
23. Housekeeping Station	23. 客房部
24. Italian Restaurant	24. 意大利餐厅
25. Children Park	25. 儿童公园
26. Japanese Restaurant	26. 日式餐厅
27. Araamu Spa	27. 阿拉姆水疗中心
28. Beach Bungalow	28. 沙滩屋
29. Water Villa	29. 水上别墅
30. Haven Villa	30. 天堂别墅
31. Haven Suite	31. 天堂套房
32. Ocean Suite	32. 海洋套房

11

5~8. Water villas
　　水上别墅
9. Outdoor dining area
　　户外就餐区
10. Beach view
　　海滩景色
11. Outdoor dining area
　　户外就餐区
12. Spa
　　水疗中心
13. Spa reception
　　水疗中心接待处

14、15. Lobby lounge
　　　大堂吧
16. Ocean suite
　　海洋套房
17. Suite relax area
　　套房休息区
18. Suite bedroom
　　套房卧室
19. Suite terrace
　　套房露台

天堂岛水疗度假村坐落于马尔代夫马累北环礁。这座私人岛屿长931米，宽201米，四周环绕着美丽的金沙和迷人的海域。天堂岛水疗度假村覆盖着大量的热带植物，彰显异域风情的海景与沙滩，为游客提供奢华之旅。

豪华的接待大厅风格独特，是休闲放松的好地方，也是查询了解情况的地方。度假村里铺设的小路穿过植被，形成一个连接网将奢华客房、各种特色餐厅、运动区域和众多服务设施连接起来，让人享受一个更加完整的假期。天堂岛水疗度假村这个自然而美丽人间天堂为游客提供了各种类型的居住设施。有220间环绕小岛一周的高级海滨小屋，还有40套水上别墅，16套天堂别墅，4间天堂套房和2间海滨套房，建在环礁湖碧蓝海水中的木桩上，极具特色并且极其舒适与安逸。每间宝塔风格的高级海滨小屋都面向海滩，离羊绒一样柔软的海岸只有几步之遥。别墅与套房的装饰现代而别致，每间套房都配有私人游泳池、极可意水流按摩浴缸和户外餐饮区。

Araamu水疗中心位于小岛隐蔽的角落，有17间用来按摩、身体护理与面部护理的特色理疗室，其中4间为专门的阿育吠陀按摩室，还有一间专门的水疗按摩室。每间私人理疗室都设有一个户外庭院、淋浴，还有花瓣浴，每一种水疗都会让你感觉置身于天堂圣地。

Sun Island Resort & Spa

太阳岛水疗度假村

Completion date: 2008 (renovation)
Location: South Ari Atoll, Maldives
Designer: Tekton Design Associate Pvt. Ltd.
Photographer: Villa Hotels & Resort
Area: 704, 000sqm

竣工时间：2008年（翻新）
项目地点：马尔代夫，南阿里环礁
设计师：Tekton联合设计公司
摄影师：维拉酒店与度假村集团
项目面积：704, 000平方米

Sun Island is the biggest resort in the Maldives and is located at the southern tip of the South Ari Atoll. Escape to an exotic world of breathtaking possibility, where adventure and relaxation are equally within reach. Located along the islands' beautiful beaches are the choice of beachfront super deluxe, deluxe or over water bungalow rooms. Its over water bungalows are spectacular in design, richly chosen and enhanced by striking ocean views that captivates the soul.

There are 136 Standard Beach Bungalows, 218 elegant Superior Beach Bungalows with its private own terrace and view of Indian Ocean. 64 Water Bungalows are

built on stilts protruding out of the turquoise shallow water in the lagoon. The 4 presidential suites are built at the far end of the water bungalow jetty. Each unit is made up of 2 bedrooms with en suite bathrooms, living room with private bar and a large private terrace with a Jacuzzi.

The Maaniyaa Restaurant is the main restaurant for full board, half board and all-inclusive guests. ANI Coffee Shop is located at the beachfront under the palm trees overlooking the sea. Sun Star Restaurant (Thai Specialty) is located on the house reef at the end of water bungalows. Southern Star Restaurant serves Buffet meals for guests accommodated in the water bungalows. The Italian restaurant & bar (Ristorente Al Pontile) with Japanese corner built on stilts over the house reef serves Italian specialties & Japanese delights. Mekunu Bar is the main bar with an entire range of liquors, cocktails and a variety of drinks. Beach bar (Sunrise Bar) is a convenient bar located near the water sports providing fruit juice and alcoholic drinks round the clock. Golf Bar views the spectacular greenery in the heart of the island. Lobby Bar is a quiet and comfortable setting for special meetings and discussions.

1. Lobby
2. Swimming pool
3. 'Vani' Coffee Shop
4. 'Maaniya' Restaurant
5. 'Mekunu Bar'
6. Southern Star Restaurant
7. 'Guraa Muli' Grill terrace
8. Golf bar & putting green
9. 'Sun Star' – Thai restaurant Shark feeding
10. Alpontile – Italian restaurant
11. Dive center – villa diving
12. Spa 'Araamu Spa'
13. Sports & recreation
14. Basketball & volleyball court
15. Tennis court
16. Soccer field
17. Medical centre
18. Sun Rise Bar & water sports
29. Sting ray feeding (sting ray show)
20. Knowledge village (villa college) / hyperbaric chamber
21. Evacuation point / assembly point
22. Nature park
23. Hydroponics
24. Retreat
25. Garden Front Bungalows
26. Standard Beach Bungalows
27. Superior Beach Bungalows
28. Water Bungalows
29. Presidential Suites

1. 大堂
2. 游泳池
3. Vani咖啡店
4. Maaniya餐厅
5. Mekunu酒吧
6. 南星餐厅
7. Guraa Muli烧烤露台
8. 高尔夫酒吧与轻击区
9. 太阳星餐厅—泰式餐厅 鲨鱼喂食区
10. Alpontile餐厅—意式餐厅
11. 潜水中心—别墅潜水
12. Araamu水疗中心
13. 运动与娱乐中心
14. 篮球场与排球场
15. 网球场
16. 足球场
17. 医疗中心
18. 日出酒吧与水上运动中心
19. 魔鬼鱼喂食区（魔鬼鱼表演）
20. 知识村（别墅学院）高压舱
21. 疏散区/集合区
22. 自然公园
23. 无土栽培区
24. 休息寓所
25. 花园景观别墅
26. 标准海滨别墅
27. 高级海滨别墅
28. 水上屋
29. 总统套房

Relax the body and awaken the senses at Araamu Spa, a sanctuary of undisturbed tranquility within Sun Island Resort & Spa. Fall under the soothing spell of traditional Maldivian healing techniques, enhanced by the natural essence of flowers, marine plants, and herbs. Beautify the body from the outside with therapeutic massages, facials and skin treatments.

太阳岛水疗度假村位于南阿里环礁最南端，是马尔代夫最大的度假胜地。这里是充满异国情调的世界，可以感受各种惊奇与冒险，又可以放松休闲。度假村有各种特色客房，包括高级奢华客房，奢华客房，还有水上屋客房，这些房屋均环岛而建，位于美丽的海滩上。水上屋的设计精美而壮观，拥有摄人心魂的迷人海景，是游客们的热选客房。

度假村建有136间标准海滨小屋，218间典雅的高级海滨小屋，设有私人露台并且可以观赏印度洋风景。64间水上屋吊脚建在环礁湖碧蓝的浅滩海水上。4套总统套房建在水上屋最远端，每个单元设有两间带浴室的卧室，带有私人酒吧间的起居室，还有带极可意水流按摩浴缸的私人大露台。

Maaniyaa餐厅是度假村的主餐厅，为全食宿游客、半食宿游客和全套服务的游客提供饮食。ANI咖啡厅面朝大海，掩映于海滨棕榈树间。太阳星泰国特色餐厅位于水上屋尽头的环岛珊瑚礁上。南方之星餐厅为水上屋中的客人提供自助餐。意大利式餐吧(Ristorente Al Pontile)与日式餐厅吊脚建在环岛珊瑚礁上，提供意大利与日本特色美食。Mekunu酒吧是度假村的主酒吧，提供各种烈酒、鸡尾酒和特色饮品。海滨日出酒吧位于水上运动附近，是提供果汁和酒精饮料的昼夜便利酒吧。高尔夫酒吧位于小岛中央，在这里可以观赏郁郁葱葱的美丽景色。大堂吧是一个宁静而舒适的特别集会场所。

Araamu水疗中心是太阳岛水疗度假村里宁静的避世胜地，在这里可以放松身体，唤醒感官神经。在自然花香精华、海洋植物，还有药草的作用下，使用马尔代夫的传统疗愈技巧，令人神经舒缓，安然入睡。通过保健按摩、面部护理和皮肤护理可以从外部作用健美体魄。

1. Resort bird view
   度假村鸟瞰图
2. Resort plan
   度假村平面图
3. Water Bungalows
   水上屋
4. Beach bar
   海滩酒吧
5. Water restaurant
   水上餐厅
6. Beach bar
   海滩酒吧
7、8. Resort main bar
   度假村主酒吧
9、10. Spa centre
   水疗中心
11. Spa massage room
   水疗按摩室

18

12. Presidential Suite
    总统套房
13. Presidential Suite bedroom
    总统套房卧室
14. Superior Deluxe Room
    高级奢华客房
15~17. Presidential Suite bathroom
    总统套房浴室
18. Water Bungalow
    水上屋

Jumeirah Dhevanafushi Maldives

马尔代夫卓美亚德瓦纳芙希度假酒店

Completion date: 2011
Location: Gaafu Alifu Atoll, Maldives
Designer: Bangkok-based Blink Design Group
Photographer: Jumeirah Dhevanafushi Maldives
Area: 44,000sqm (total area of the uninhabited
island plus a unique 'water village')

竣工时间：2011年
项目地点：马尔代夫，卡夫阿里夫环礁
设计师：曼谷布林克设计集团
摄影师：马尔代夫卓美亚德瓦纳芙希度假酒店
项目面积：44,000平方米（无人岛与特别水村总面积）

The resort has a unique two island concept, with the main island occupying an area of 44,000sqm and the 'Ocean Pearls', an exclusive collection of overwater villas, located 850metres from the main island, right in the heart of the ocean and surrounded by pure coral reefs. The design of the resort was done by Bangkok-based Blink Design Group, inspired by Sri Lankan architect Geoffrey Bawa, one of the most important and influential Asian architects of the twentieth century.

The vision of the designers is to create an exclusive 'two-resort island' reflecting boutique escape luxury, private island spirit, haute couture, discreet and glamorous

ambience, personalised and thoughtful service, privacy and exclusivity. Translating into 'unique island', Dhevanafushi takes the essence of the Maldives to create new and distinctive experiences for the guests. The uniqueness of every guest is reflected in the originality and space of the interior design.

At the resort, there are two principle residence types, Revive and Sanctuary, each offering a choice of breathtaking views over island or ocean. Magnificently designed, with honey coloured timber flooring and soaring roofs, every abundantly sized room is inspired by traditional Maldivian architectural expressions. 13

beautiful Beach Revives are perfectly positioned by the ocean, while seven Island Revives lie nestled amidst areas of rich flora, creating a distinctive residential feel. The unique and completely secluded Island Sanctuary is expected to open 2013.

Spacious bedrooms with timber flooring and high ceilings include a separate study area, decorated with neutral colour palettes, infusing peace and calm; while a spacious open bathroom and shower create a space of graceful elegance. Guests are spoilt with their own swimming pool, private beach access and gardens, and private sala day beds on their private deck.

A short boat ride away from the main island, Jumeirah Dhevanafushi's 'Ocean Pearls' is laid out for the most discerning luxury traveller. With 14 Ocean Revives and two Ocean Sanctuaries, guests can spend unforgettable days and enchanted evenings overlooking the Indian Ocean, experiencing a unique level of remote luxury from the comfort of their luxurious and beautifully appointed suite.

Built above the crystalline waters of the Indian Ocean at the resort's Ocean Pearls, each of these 14 beautiful Ocean Revives offers an elevated sense of luxury with soothing colours and capacious interiors, perfectly complemented with large floor to ceiling windows in all rooms including the bathroom. The beds – 3x2 metres and the largest in the Maldives – offer an unrivalled level of comfort. The private deck with a marble round bathtub and infinity-edged pool offers sublime ocean views. Steps from your deck lead directly into the lagoon and ocean.

With stunning views of Maldivian sunsets you are surrounded by beauty and luxury. With generous interiors, traditional Maldivian designs, private decks and indulgent marble baths, this is the epitome of everything exquisite. Ocean Sanctuaries have two bedrooms, a separate living and dining area, study, large panoramic windows that offer unrivalled views.

The resort features three restaurants and one bar, offering a variety of culinary options. At Jumeirah Dhevanafushi, our philosophy is based on recognising individual needs and we have developed a unique personalised gastronomic experience. The spaces across all restaurants are designed to delight in the incredible unspoiled views of the Ocean.

Talise spa covers 1,700sqm and includes 3 private over-water treatment rooms and one VIP room, hammam, steam, sauna, incorporated in the treatment rooms. Talise Spa specialises in treatment rooms for couples and is set in luxurious surroundings which include a yoga platform over water and a viewing jetty to sit and enjoy the breathtaking sunsets. Jumeirah Dhevanafushi offers a 190sqm swimming pool, a 70sqm fully-equipped over-water gym, a Yoga Pavilion and an extensive selection of land and water sports activities.

1. Arrival jetty
2. Library reception
3. Khibar
4. Infinity pool
5. Azara
6. Watersports centre
7. Gym
8. Talise Spa
9. Talise Spa lounge
10. Talise Spa reception
11. Island Sanctuary
12. Yoga platform
13. Mumayaz
14. Beach revives
15. Island revives
16. Diving centre

1. 到达码头
2. 图书室前台
3. Khibar酒吧
4. 无边界泳池
5. Azara主题餐厅
6. 水上运动中心
7. 健身房
8. 泰丽丝水疗中心
9. 泰丽丝水疗休息室
10. 泰丽丝水疗前台
11. "岛屿隐世圣殿"别墅
12. 瑜伽平台
13. Mumayaz休闲海滩烧烤餐厅
14. "海滩活力空间"别墅
15. "岛屿活力空间"别墅
16. 潜水中心

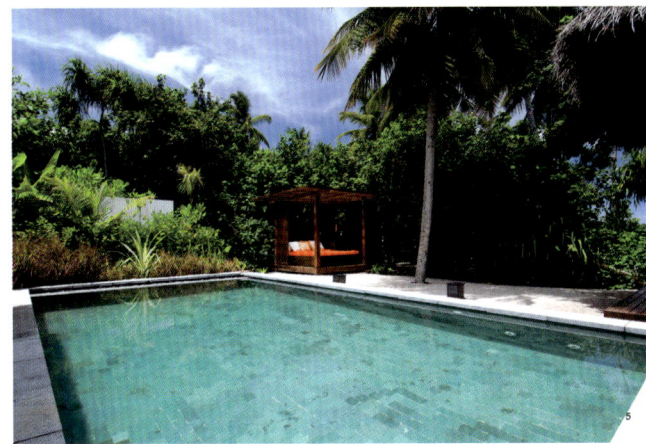

该度假酒店采用独特的双子岛概念，其主岛面积为44,00平方米，距离主岛850米处的海中央，有一座被天然珊瑚礁环绕着的水上别墅，被称为"海洋珍珠"。 度假酒店由曼谷的布林克设计集团（Blink Design Group）负责设计，灵感源自斯里兰卡著名建筑师、同时也是20世纪最具影响力的亚洲建筑师——杰弗里·巴瓦(Geoffrey Bawa)。

设计者的用意是精心打造一座精致、隐秘、奢华和富有私人情调的"双度假岛屿"，岛上建有精品别墅、高级时装店和各种私人娱乐休闲设施，提供独特私密的个性化、细致服务。德瓦纳芙希（Dhevanafushi）别指"独特的岛屿"。卓美亚德瓦纳芙希度假酒店汲取了马尔代夫的精华，能够为游客带来极具特色的全新体验。酒店深刻意识到每位高端游客的个性化需求以及个人体验的重要性，而其内部设计的原创性和空间感则充分体现出酒店对每位游客独特性的高度重视。

该度假酒店有两种类型的别墅，即"活力空间（Revive）"和"隐世圣殿(Sanctuary)"，两者均拥有令人惊叹的岛屿风光或海上风光。别墅设计华丽，蜜色的实木地板、高挑的屋顶和宽敞的房间无不体现出马尔代夫的传统建筑风格。13幢"海滩活力空间"（Beach Revives）别墅面朝印度洋，7幢"岛屿活力空间"（Island Revives）别墅依偎在茂密的丛林中，给人一种与众不同的居住感受。还有一幢独特的"岛屿隐世圣殿"（Island Sanctuary）别墅有望在2013年问世。

宽大的卧室以实木地板装饰，屋顶高悬。配有一个独立的阅读区，中性色调装饰给人一种平静安宁的感觉；明亮宽敞的户外浴室和花洒淋浴，营造出独特的优雅空间。客人们可以享受他们的私人泳池和海滩、花园，抑或躺在其专属的露天阳台长椅上享受日光浴。

乘船从主岛出发，几分钟即可抵达卓美亚德瓦纳芙希度假酒店的水上度假村"海洋珍珠"，它专为满足最挑剔的高端游客需求而设计。度假村有14幢"海洋活力空间"别墅和2幢"海洋隐世圣殿"。客人在此将度过令人难忘的休闲假日，俯瞰印度洋迷人的夜景，感受奢华别墅和美丽套房之外的独特奢华体验。

这14幢精美绝伦、坐落在印度洋碧蓝水域上的"海洋珍珠"别墅，每幢均拥有柔和的色调和宽敞的室内空间，给人以无比奢华的感觉。包括浴室在内，所有房间都设有从地板直到天花板的落地窗，而别墅3米×2米的大床为马尔代夫最大，提供了无与伦比的舒适享受。私人阳台上配有大理石圆形浴缸和巨大的泳池，提供了恢弘的海景，从阳台还能直接步入环礁湖和大海。

马尔代夫的日落美不胜收，令人惊叹；别墅设计奢华优雅，内饰宽敞大方；传统的马尔代夫设计、私人隐秘的露台和无比舒适的大理石浴缸，这里尽显精细设计理念。

"海洋隐世圣殿"有两个卧室，起居室和用餐区分隔开，书房以及全景落地窗，皆提供了无与伦比的美景。

度假酒店有3个餐厅和1间酒吧，提供各式美食佳酿。卓美亚德瓦纳芙希度假酒店的理念是满足每一位客人的需求，为客人们提供独具特色的个性化美食体验。每个餐厅的设计都保证客人能观赏到难以置信的无敌海景。

泰丽丝水疗中心占地1,700平方米，拥有3间水上理疗室和1间贵宾理疗室，理疗室内设有土耳其浴室、普通蒸汽浴室和芬兰式蒸汽浴室。泰丽丝水疗中心专为情侣夫妇提供配置豪华的理疗室，其中包括一个海景瑜伽区和一个观景台，供客人观赏迷人的日落景色。卓美亚德瓦纳芙希度假酒店拥有一个190平方米的游泳池、一间70平方米配套齐全的海景健身房和一个瑜伽馆，同时还提供各种陆地和水上运动项目。

Constance Moofushi Resort

康斯坦斯慕芙岛度假村

Completion Date: 2010
Location: South Ari Atoll, Maldives
Designer: Studio Marc Hertrich & Nicholas Adnet
Photographer: Studio Marc Hertrich & Nicholas Adnet
Area: 68, 000sqm

竣工时间：2010年
项目地点：马尔代夫，南阿里环礁
设计师：Marc Hertrich & Nicholas Adnet设计工作室
摄影师：Marc Hertrich & Nicholas Adnet设计工作室
项目面积：68,000平方米

Constance Moofushi Resort is situated on the South Ari Atoll. The Resort combines the Crusoe Chic Barefoot elegance of a deluxe resort with the highest standards of Constance Hotels Experience. Based on an array of beautiful scenery, the Moofushi island resort in the Maldives is the perfect place for a relaxing holiday.

Studio MHNA has designed a pearl in the middle of the Maldives archipelago, with the beauty of the ocean as the background. The whole concept is luxurious, but the project has been developed with a true desire of not showing off. Here, lifestyle blends well with wild beauty. The range of colours is purposely neutral and natural.

It is a monochrome of white, beige, greige, ivory and sand punctuated with bright colours from the seabed and wood which plays an important part in the whole design.

The amazing retreat welcomes its guests with a surprising collection of inside and outside spaces. Featuring 24 Beach Villas, 56 Water Villas, and 30 Senior Water Villas, the retreat complex is one of the most luxurious resorts in South Ari Atoll in Maldives. Offering tranquillity and relaxation, the resort had to be constructed to take advantage of tropical climate conditions: rainwater harvesting and waste treatment plants were implemented; special green design elements include high roofed areas and open ceilings, cross ventilation in all indoor spaces, deep roof overhangs, and window shading.

The over-water extension of the restaurant, allows guests to enjoy a perfect view on the crystal-clear waters of the Indian Ocean. Located a few metres from the sea on white sandy beaches, the beach grill features a casual and relaxed atmosphere. Luxurious but unique design also goes to the spas, the spa villas, the Yoga Pavilion and other entertainment facilities such as the Swimming Pool, Gym & Acrobic and Entertainment Lounge.

1. Senior Water Villas
2. Water Villas
3. Beach Villas
4. Boathouse
5. Gym
6. Totem Bar
7. Beach Grill
8. Spa De Constance
9. Spa Reception

10. Boutique
11. Diving Centre
12. Reception
13. Manta Bar
14. Manta Restaurant
15. The Deck
16. Entertainment Lounge
17. Guest Arrival Jetty

1. 高级水上别墅
2. 水上别墅
3. 海滨别墅
4. 船屋
5. 健身中心
6. 图腾酒吧
7. 海滨烧烤区
8. 康斯坦斯水疗中心
9. 水疗中心接待处

10. 精品店
11. 潜水中心
12. 接待处
13. 曼塔酒吧
14. 曼塔餐厅
15. 甲板
16. 娱乐休闲吧
17. 客人到达码头

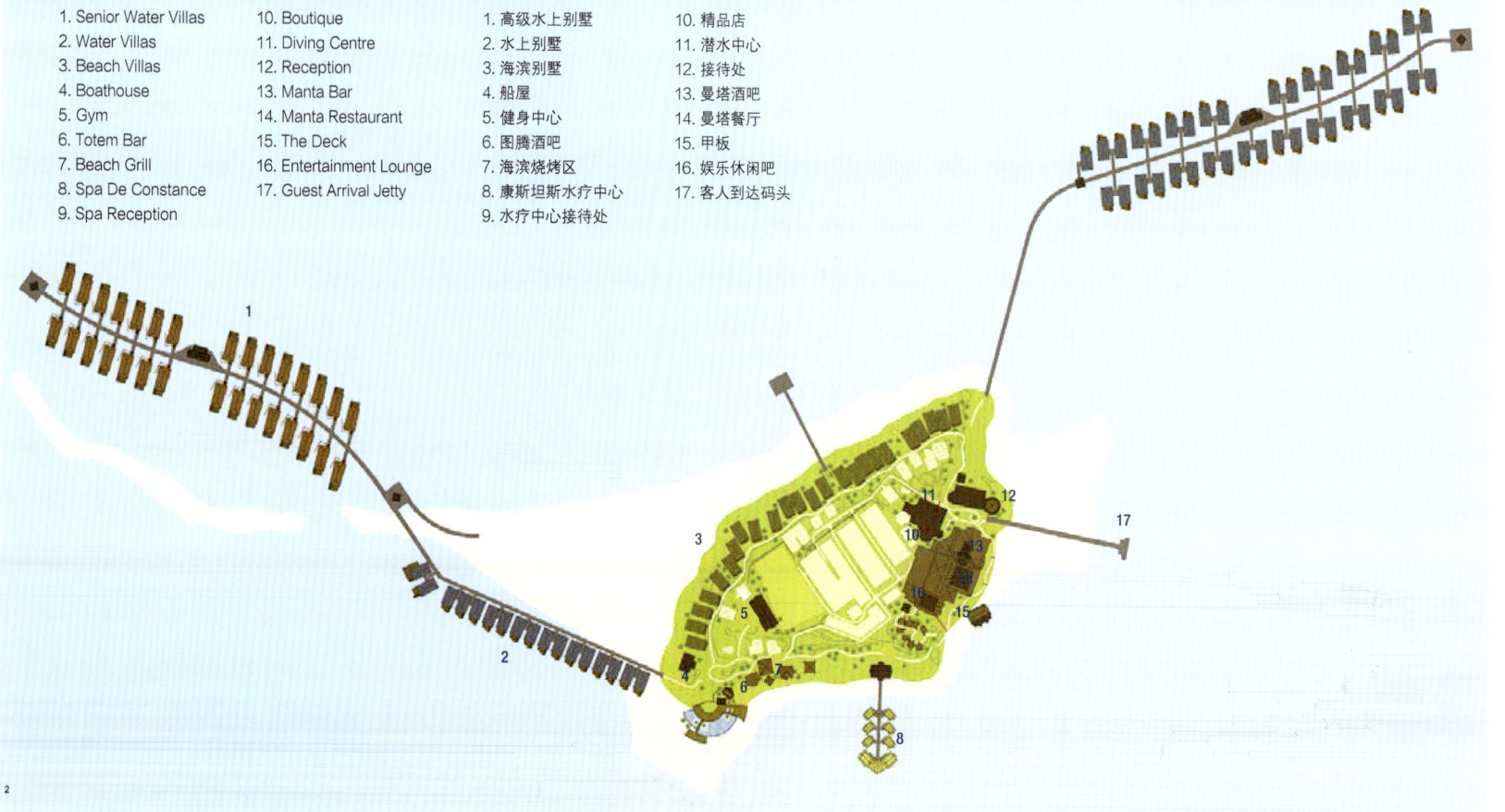

康斯坦斯慕芙岛度假村坐落在南阿里环礁上，是一个将克鲁索式的别致典雅与康斯坦斯酒店最高标准结合在一起的豪华度假村。位于马尔代夫的慕芙岛度假村风景优美，是休闲度假的完美之地。

MHNA设计工作室以美丽的海洋为背景，在马尔代夫群岛的中心设计了一颗明珠。项目的整体设计理念是奢华，但其实在设计过程中又在追求低调，使传统的生活方式与旷野的美丽相融合。设计中特意使用了中性色与自然色，整体设计中使用了纯白色、米黄色、乳白色和沙白色这样的单色，又以海床和木材的亮色来增强突出的效果。

这个令人惊艳的避世胜地无论是室内还是室外都给游客提供了广阔的空间。该度假村建有24幢海滨别墅，56幢水上别墅，30幢高级水上别墅，是马尔代夫南阿里环礁地区最豪华的度假村之一。为了营造宁静而放松的氛围，该度假村的建造充分利用了该地区的热带气候条件，采用了集水技术和污水处理措施；使用了特殊的绿色环保的设计元素，包括深屋顶和开放式天花板的设计，室内空间对流通风，深屋顶悬臂结构，还有遮阳窗。

在延伸的水上餐厅里，游客可以欣赏印度洋水晶般清澈的海水与美丽的景色。海滩烧烤区距大海只有几米，营造了一种休闲放松的氛围。奢华而独特的设计也运用到了水疗中心、水疗别墅、瑜伽亭和游泳池、健身中心、休闲吧等其他娱乐设施中。

1. Totem Bar
   图腾酒吧
2. Resort plan
   度假村平面图
3. Resort pool view
   度假村泳池景色
4. Beach view
   海滩景色
5. Villa terrace
   别墅露台

6. Overwater spa exterior
   水上水疗中心外观
7. Overwater spa terrace
   水疗中心露台
8、9. Spa treatment room
   水疗中心治疗室

Kunuhura, Hotel in Maldives

马尔代夫卡努呼拉岛度假酒店

Completion date: 2005(renovation)
Location: Lhaviyani Atoll, Maldives
Designer: Tekton Design Associates
Photographer: Kanuhura, Hotel in Maldives
Area: 300,000sqm

竣工时间：2005年（翻新）
项目地点：马尔代夫，拉维亚薇环礁
设计师：Tekton联合设计公司
摄影师：马尔代夫卡努呼拉岛度假酒店
项目面积：300,000平方米

Kanuhura is located on the eastern rim of Lhaviyani Atoll, on its own private island. Every effort has been made to retain the natural beauty of the island and wherever possible, only natural materials are employed. The casual elegance of the island is preserved with the choice of natural materials and a colour palette pulled from the sea.

Kanuhura encompasses 100 villas in five categories, all with sea views and either access to their own white sand beach or a private wooden staircase leading straight down into the clear waters of the lagoon. The overwhelming impression as guests enter a villa is of a

retreat from the sun, the sand and the sea; and yet the view through the floor-to-ceiling sliding glass doors is always out to the mesmerising ocean. Extensive use of hand-crafted wood, raffia, bamboo and other natural materials creates an elegant rustic interior befitting the naturally beautiful island surroundings. The plantation shutters at the windows, while wardrobes, furnishings and bed are in solid hand-crafted wood, with raffia or split bamboo finishes, and honed natural grey slate surfaces in the dressing room area.

The king-size customised bed is draped with sheer fabric overhead and bed linen of the finest Egyptian cotton; walls are plain and cool; soft furnishings are of linen and slub cotton in earth tones with splashes of hot tropical colours; lights are dimmed with sand-coloured shades and the walls are decorated with themed prints. Ceiling fans and individually controlled air-conditioning allow guests to control the room climate.

Kanuhura resort include three distinctive restaurants and one bar. Thin Rah is a large airy restaurant with its fold-back louvre windows, immense thatched roof with inverted funnels letting in light and air, and outdoor eating area with sand floor and view of

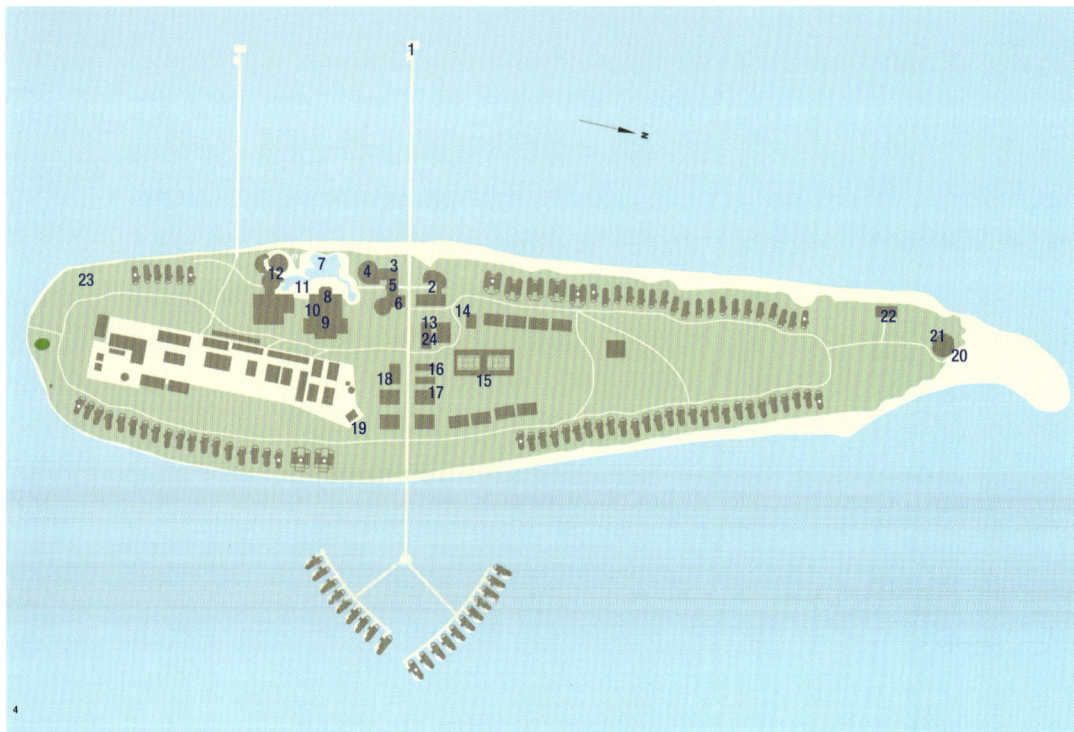

1. Arrival Jetty and Moodhu Lounge	1. 到达码头与穆德胡休闲吧
2. Reception	2. 接待处
3. Lava Lounge	3. 拉瓦休闲吧
4. Handhuvaru Bar	4. 汉德胡瓦如酒吧
5. Games room	5. 游戏室
6. Lafuz Library	6. 拉弗兹图书室
7. Swimming pool	7. 游泳池
8. Spa	8. 水疗中心
9. Fitness Studio	9. 健身室
10. Gymnasium	10. 体育馆
11. Olive Tree	11. 橄榄树
12. Thin Rah (Main restaurant)	12. 馨拉餐厅（主餐厅）
13. Sun Dive Centre	13. 阳光潜水中心
14. Hospitality room	14. 迎宾室
15. Tennis courts	15. 网球场
16. Squash court	16. 壁球场
17. Jewellery shop	17. 珠宝店
18. The boutique	18. 精品店
19. Mosque	19. 清镇寺
20. Veli Café	20. 微莉咖啡厅
21. Nashaa Club	21. 纳莎俱乐部
22. Water sports	22. 水上运动中心
23. Kids club	23. 儿童俱乐部
24. The clinic	24. 诊所

the main pool and lagoon. Thin Rah has three open-sided indoor pavilions. All areas feature hand-blown glass screen sculptures under peaked thatched roofs. Poolside open-air tables and a breezy interior offer shady refuge by day, while in the evening an intimate atmosphere is created when the lights dim and the candlelight glimmers. Candlelit romance on the shores of the Indian Ocean sets the tone for Veli Café, with its elegantly decorated timber tables and chairs set amidst the palm trees on the beachside wooden terrace. The spacious, partly sand-floored Handhuvaru Bar on the other side of the pool from the two main restaurants has been dressed with comfortable over-sized daybeds and chairs both indoors and outdoors, by the pool and on the beachfront. The interior is with inviting, cosy seating and windows wide-open to the daylight.

At Spa, timber flooring, complementary lighting concepts and a fresh colour palette for walls and fabrics offset daybed furnishings in the relaxation room. Everything about this Asian-style spa contributes to a sense of tranquillity: aromatic scents, surroundings of warm brown wood, billowing cotton on the ceiling, carved stone wall panels, ornamental lily ponds, bowls of water with floating frangipani petals, daybeds with cotton turndowns, shady thatched roofs which admit natural daylight and a light sea breeze. The spa comprises eight treatment rooms, including four doubles for side-by-side couples' treatments, as well as an outdoor beachside spa pavilion, Kandu-Olhi.

1. Beachside water villa
   海滨水上别墅
2. Resort sea view
   度假村海景
3. Water villa
   水上别墅
4. Resort plan
   度假村平面图

5、6. Resort reception area
   度假村接待区
7. Beach villa
   海滨别墅
8. Private retreat with pool
   私人泳池别墅
9. Grand beach villa
   高级海滨别墅

10. Olive tree restaurant
   橄榄树餐厅
11. Private retreat
   私人别墅
12. Veli café
   薇丽咖啡厅
13. Handhuvaru bar
   Handhuvaru酒吧

14. Moodhu lounge
   Moodhu休闲吧
15. Thin-rah beach restaurant
   Thin-rah海滩餐厅

卡努呼拉岛度假村位于拉维亚薇环礁东端一个私人岛屿上。设计中竭力保持岛上的自然美景，并且尽可能地使用天然材料。天然材料的使用，和来自海洋的配色方案保持了小岛休闲典雅的氛围。

卡努呼拉岛度假村共有100幢别墅，分为5个类别，每幢别墅都能欣赏到美丽的海景，还可以通往白沙滩或者通过一个私人木制楼梯直接通往环礁湖清澈的海水中。游客进入别墅最深刻的感觉就是来到一个充满阳光、沙滩和海水，还可以通过落地玻璃滑门观赏迷人海景的避世胜地。设计师广泛使用了手工木、拉菲亚木、竹子和其他天然材料，创造了一个与岛上的自然美景相融合的纯朴而典雅的室内空间。百叶窗、衣柜、家具和床都是由手工实木制成，还有拉菲亚木或者竹制的家具，更衣室使用了亚光面的天然灰色石板。

特大号定制床上的床单和枕套由精品埃及棉制成，床头悬挂着透明织物，墙面朴实而清爽；室内以亚麻和竹节纱装饰，主色调为大地色，并且点缀着强烈的热带色；灯具使用了渐暗色调，墙面以主题壁纸装饰。客人可使用天花板吊扇和独立调控中央空调自由调控室内温度。

卡努呼拉岛度假村有三个特色餐厅和一个酒吧。Thin Rah餐厅是一个大的通风餐厅，使用了可折叠式百叶窗；巨大的茅草屋顶中间是一个倒漏斗式的设计，可以射入自然光，流入清新的空气；还有户外餐饮区，这里有沙床，还可以观赏主泳池及环礁湖美景。Thin Rah餐厅有三个开放式的室内亭，以尖形茅草屋顶下人工吹制的玻璃幕雕塑为特色。泳池边设有露天餐桌，还有室内餐厅，白天荫凉通风，晚上灯光昏暗，烛光微耀，营造了一种亲密的氛围。Veli咖啡厅装饰典雅，木制桌椅放置在海滨木制露台上，掩映在海滨木制露台棕榈树中，在微弱的烛光下营造出印度洋海岸的浪漫氛围。泳池的另一侧是两大主餐厅中的Handhuvaru酒吧，空间宽敞，部分位于沙床上，室内和室外的泳池边和海滩上都装饰有舒适的超大躺椅和扶手椅。室内放置了舒适诱人的座椅，通过完全开放的窗户可以接受日光的照射。

水疗中心里使用了实木地板和现代理念的照明，墙面和休息区中躺椅上的织物色彩清新。水疗中心采用了亚洲风格的装饰，这里的一切都是为了营造一种宁静的氛围，包括芳香的气味，四周暖棕色的木材，波浪形的天花板，石刻护墙板，装饰性的莲花池，漂浮着赤素馨花的水碗，铺有棉垫的躺椅，还有可以接收自然光和轻柔海风的茅草屋顶。水疗中心设有8间治疗室，其中包括四间夫妻并排治疗室，和一个名为Kandu-Olhi的户外海滨水疗亭。

16. Spa lobby lounge
    水疗中心大堂吧
17. Spa pool
    水疗池
18. Spa relaxation area
    水疗中心休息区
19. Beach villa terrace
    海滨别墅露台
20. Grand beach villa living room
    高级海滨别墅客厅

21. Grand beach villa bedroom
    高级海滨别墅客房
22. Private retreat bedroom
    私人别墅客房
23. Grand beach villa couple bedroom
    高级海滨别墅双人间客房
24. Private retreat bathroom
    私人别墅浴室

Matahari Beach Resort & Spa

玛塔哈莉海滨水疗度假村

Completion date: 2009 (renovation)
Location: Bali, Indonesia
Designer: IR. Komang Suardana, Parwathi and Magnus Bauch
Photographer: Matahari Beach Resort & Spa
Area: 40, 000sqm

竣工时间：2009年（翻新）
项目地点：印度尼西亚，巴厘岛
设计师：艾尔康芒·苏尔达纳，帕瓦提和马格纳斯力·鲍克
摄影师：玛塔哈莉海滨水疗度假村
项目面积：40,000平方米

Nestled between the Bali Barat National Park and the Java Sea, the secluded and exclusive Matahari Beach Resort & Spa is the perfect destination for connoisseurs of luxury and relaxation. You can wake up with the bird songs in the tropical garden, listen to the ocean-waves during your morning yoga session, explore quaint neighbourhood villages and surroundings on a bike and be pampered in the award-winning Spa before you indulge yourself with a gourmet dinner in a romantic pavilion.

For accommodation, there are four room types: Garden View, Premium Garden View, Deluxe and

Super Deluxe. Using local materials like sandstone and bamboo, local craftsmen and artists handcrafted the bungalows that blend harmoniously with the tropical garden.

The Dewi Ramona Restaurant enjoys a reputation for providing exceptional culinary experiences. You can either choose from the many specialties on the extensive a la carte menu that changes daily or enjoy the informal, blissful atmosphere of Leon's Beach Bistro. Meals are prepared using the freshest ingredients, harvested from organic farms and plantations. Wine out of the impressive wine cellar will complement every meal.

Built in the style of an ancient Balinese royal palace, the award-winning Parwathi Spa is a sanctuary of relaxation. Massage techniques and aroma oils are used to revitalise and rejuvenate your body, mind and soul.

玛塔哈莉海滨水疗度假村坐落于巴厘岛巴拉特国家公园和爪哇海之间，避世而独特，是追求奢华与休闲之旅的最佳胜地。伴随着热带花园里的鸟鸣声醒来，清晨瑜伽时可以倾听海浪声，可以骑着单车去探索周围的村庄和一些稀奇古怪的事物，可以在浪漫的凉亭中享用美食，还可以在用餐前在水疗中心尽情放松。

客房分为四种类型：花园景观客房、高级花园景观客房、奢华客房和超级奢华客房。这些别墅小屋均使用当地材料，如砂岩和竹子，由当地手工匠和设计师们亲手打造而成，与当地的热带花园和谐的融合在一起。

简廷芮拉蒙纳餐厅以提供特色餐饮体验而享有盛名。利昂斯海滨酒馆气氛悠闲，充满喜悦，在这里可以从每日更换的菜单中选择各种特色美食。美食由来自当地农场与种植园的最新鲜的材料烹饪而成，并且每餐提供的葡萄酒都是来自特色酒窖，令人印象深刻。

备受赞誉的帕瓦提水疗中心以古代巴厘岛皇家宫殿风格建造而成，是休闲放松的避世之所。利用高超的按摩技术，配合芳香的精油，可以唤醒你的身体、大脑和灵魂。

1. Lobby	15. Diving school	1. 大厅	15. 潜水学校
2. Conference room	16. Beach restaurant	2. 会议室	16. 海滨餐厅
3. Dewi Ramona Restaurant	17. Crafts	3. 戴韦雷蒙娜餐厅	17. 工艺品店
4. Wayang Bar	18. Pitch & Put	4. 哇扬酒吧	18. 投掷场
5. Gallery	19. Parwathi Spa	5. 画廊	19. 帕瓦提水疗中心
6. Beauty Salon	20. Gym	6. 美容沙龙	20. 健身中心
7. Library	21. Tea pavilion	7. 图书室	21. 茶亭
8. Open Air stage	22. Yoga pavilion	8. 露天舞台	22. 瑜伽亭
9. Pool	23. House temple	9. 游泳池	23. 殿堂之屋
10. Children's pool	24. Garden View	10. 儿童泳池	24. 花园景观客房
11. Lotus pavilion	25. Premium Garden View	11. 莲花亭	25. 顶级花园景观客房
12. Beach	26. Deluxe	12. 海滩	26. 奢华客房
13. Beach Massage pavilion	27. Super Deluxe	13. 海滨按摩亭	27. 高级奢华客房
14. Sea		14. 大海	

19. Super Deluxe Bungalow bedroom
    超豪华别墅卧室
20. Garden View Bungalow bedroom
    花园景观别墅卧室
21. Super Deluxe Bungalow bathroom
    超豪华别墅浴室
22. Parwathi Spa Flowerbath
    帕瓦提水疗中心花瓣浴
23. Garden View Bungalow bathroom
    花园景观别墅浴室

Al Areen Palace & Spa

艾尔阿润宫水疗度假村

Completion date: 2009
Location: Manama, Bahrain
Designer: Al Hamad
Photographer: LHW
Area: 131, 309sqm

竣工时间：2009年
项目地点：巴林，麦纳麦
设计师：艾尔·哈马德
摄影师：LHW
项目面积：131, 309平方米

Al Areen Palace & Spa is a real life Arabian Nights fantasy, with private villas and a stunning setting overlooking the Arabian Gulf on Bahrain. Boasting the world's first garden hammam, this five-star luxurious resort offers villas with private pools in the naturally beautiful Sakhir. It combines Arabian-style design, luxurious furniture and contemporary flair.

There are three international restaurants and a bar & lounge in the resort. Vertigo Bar & Lounge offers breathtaking panoramic views of the Arabian-styled villas below.

A delectable mix of Middle East artistry and

contemporary design, Desert Pool Villa exudes the hospitality and warmth of the inhabitants on this island of golden smiles. Each Desert Pool Villa is highlighted by an en-suite bathroom and a separate living area. Elegant and airy interiors speak of modern comforts with a touch of traditional artistry.

The romance of the Arabian Nights is rekindled to perfection in the luxuriously appointed Royal Pool Villas at Al Areen Palace & Spa. Voluminous, high vaulted ceilings bring the warmth of ornate external architecture indoors where modern comforts prevail. Each of the 22 Royal Pool Villas features two sprawling bedrooms, each boasting an inviting king-size bed, complete with en suite bathrooms with oversized bathtubs. The elegant ambience extends to the plush living area fitted with a wide range of modern amenities. A key feature is a 10,000 square-metre spa, the largest in the Middle East and a world first in terms of size, design and treatments. The Spa features, twelve Spa pavilions, each with twin rainmist shower beds, infinity bathtubs and a private garden, Hydrothermal Garden and Garden Hammam, Ladies' Beauty Garden, Gentleman's Salon and extensive fitness and wellness facilities.

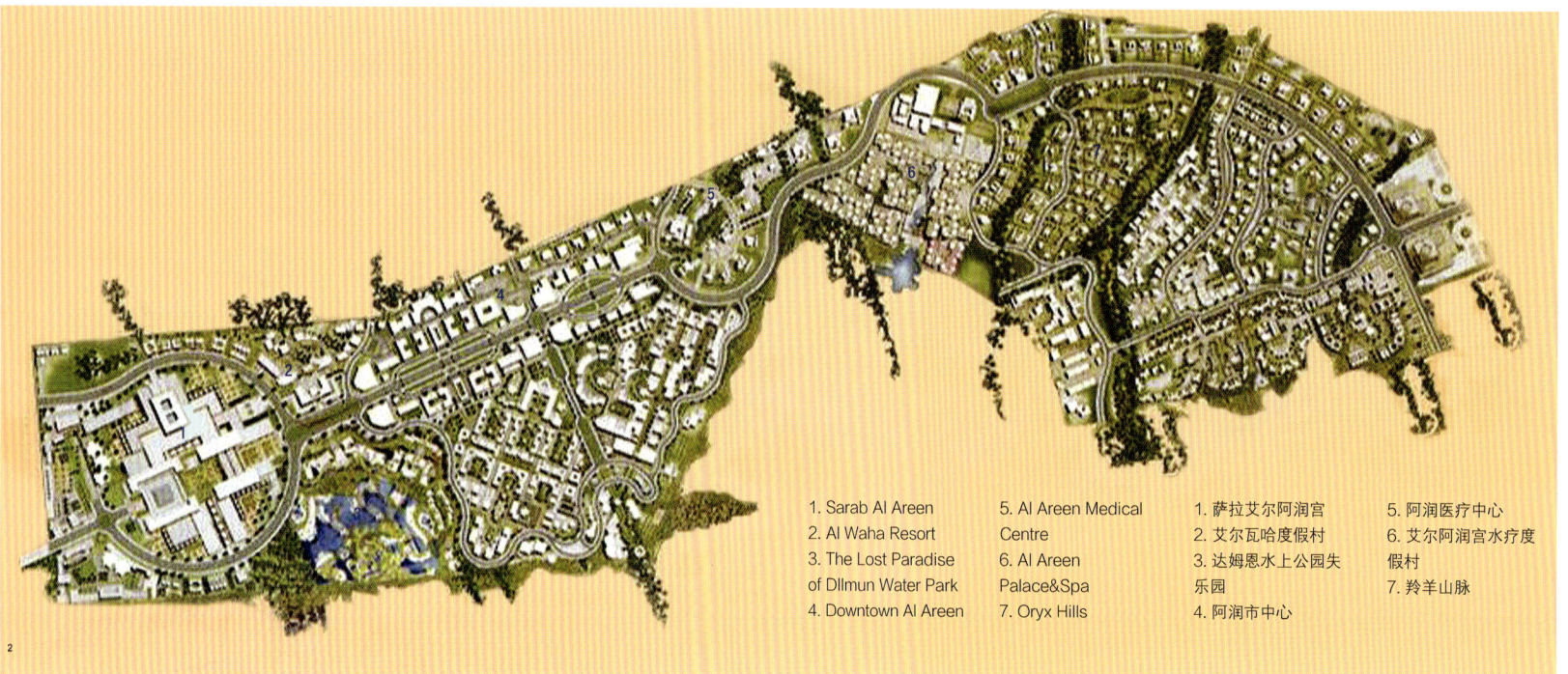

1. Sarab Al Areen
2. Al Waha Resort
3. The Lost Paradise
   of DIlmun Water Park
4. Downtown Al Areen

5. Al Areen Medical
   Centre
6. Al Areen
   Palace&Spa
7. Oryx Hills

1. 萨拉艾尔阿润宫
2. 艾尔瓦哈度假村
3. 达姆恩水上公园失
   乐园
4. 阿润市中心

5. 阿润医疗中心
6. 艾尔阿润宫水疗度
   假村
7. 羚羊山脉

艾尔阿润宫水疗度假村是一千零一夜中生活的真实体现，这里建有私人别墅，可以俯瞰阿拉伯湾的绝美景色。这是一家5星级的奢华度假村，建有世界一流的花园哈曼浴室，还有建在自然而美丽的萨基尔赛道上的私人泳池别墅。度假村的设计结合了阿拉伯式的设计风格和当代风格，并使用了奢华的家具设施。

度假村设有三间国际性的餐厅和酒吧与休闲吧。在Vertigo酒吧与休闲吧中可以俯瞰下面阿拉伯风格别墅的惊人全景。

每间沙漠泳池别墅都摆放着令人愉悦的中东艺术品，并结合了当代设计风格，散发着这个金色微笑之岛上居民的好客和热情。每间沙漠泳池别墅都设有一间套房浴室和一间单独的起居室。清新典雅的室内设计结合一些传统艺术品的装饰体现了现代的舒适。

艾尔阿润宫水疗度假村中豪华的皇家泳池别墅完美重现了一千零一夜中的浪漫氛围。大量采用高拱形天花板，使华丽的室内建筑外观更显现代的舒适与温暖。22套皇家泳池别墅中，每套都设有两间大卧室，卧室中摆放着引人注目的超大号床，别墅中还有带超大浴缸的套房浴室。这种典雅的氛围也弥漫到了带有各种现代服务设施的豪华起居室。

10,000平方米的水疗中心是该度假村的主要特色，它是中东地区最大的水疗中心，无论在规模、设计和护理方面都堪称世界一流。水疗中心有12间带双人浴床和超大浴缸的水疗亭，还有一间私人花园浴室、热水花园浴室、花园哈曼浴室及女士花园美容中心、男士沙龙和大量的美容健身设施。

1. Saffron Restaurant exterior
   Saffron餐厅室外
2. Resort plan
   度假村平面图
3、4. Royal pool villa
   皇家泳池别墅
5. Desert pool villa
   沙漠泳池别墅
6. Villa connecting walkway
   连接别墅的通道
7. In-villa dining
   别墅内餐厅

8. Restaurant
   餐厅
9. Restaurant private dining.
   餐厅私人就餐区
10. Rimal restaurant
    Rimal餐厅

11. Spa pool
水疗池
12. Villa bathroom
别墅浴室
13. Deluxe massage pavilion
豪华按摩亭

14. Spa entrance
水疗中心入口
15、16. Villa bathroom
别墅浴室

The Grand Mauritian Resort & Spa in Mauritius

毛里求斯豪华水疗度假村

Completion date: 2008
Location: Balaclava, Mauritius
Designer: Africon
Photographer: The Grand Mauritian Resort &
Spa in Mauritius
Area: 100,000sqm

竣工时间：2008年
项目地点：毛里求斯，巴拉克拉法
设计师：艾弗里孔设计公司
摄影师：毛里求斯豪华水疗度假村
项目面积：100,000平方米

The resort is located on a beautiful stretch of coastline in the North West of the island. A romantic, typical Mauritian feel was key to the design. The comfort and simplicity of the island has been embraced and transformed into a style that typifies the new contemporary Mauritius. The architecture has a laid-back tropical feeling that evokes memories of the days gone-by, whilst maintaining a sophisticated quality in keeping with the standards of a five-Star Resort.

The interiors take inspiration from the ocean and its waves with blue hues and taupe neutrals employed. German silver and other chrome accents represent the

moon and starlight reflections on the Indian Ocean, creating a cool, calm and fresh feel.

Much of the The Grand Mauritian has been built with a traditional hardwood timber called Balau, whilst the incorporation of the natural volcanic lava rock, evident throughout the island, is widely used. Immediately noticeable too is the extensive use of thatching as all the villas and main building enjoy 'sugar cane frond' roof-coverings.

The other finishes such as the Bottichino marble floor tiles have been brought into the design to add certain modern classic refinement and sophistication. The

resort consists of 193 rooms, offering a very varied room mix which includes deluxe rooms, beach club rooms and a few select suites, in which large private outdoor terraces are common.

The guest rooms are modern classical, yet stylish in design and offer all of the latest amenities that a guest would require. Soft muted colours such as taupes, ivory and duck egg blue have been used to create a sophisticated yet relaxed feel to the room. Large open plan marble bathrooms and dressing area open up onto a generous bedroom area and patio with spectacular views of the sea and the coastline with natural reef.

度假村坐落在岛屿西北面延展开的美丽海岸线上。设计的重点是浪漫而典型的毛里求斯风格。设计崇尚岛屿的舒适和简洁，并将其转化为现代新毛里求斯风格。建筑有一种慵懒的热带风情，唤醒了人们昔日的回忆，同时又保证了五星级度假村的高级品质。

室内设计从海洋和灰蓝色的波浪中得到了灵感，采用了中性的色调。德国银和其他金属色描绘出迷人的月亮和印度洋上的点点星光倒影，营造出清爽、平静、新鲜的感觉。

毛里求斯豪华水疗度假村的大部分建筑都采用了传统木材——娑婆双木，同时融入了遍布全岛的天然火山熔岩。设计还大量使用了茅草屋顶，为别墅和主要建筑添加了"甘蔗叶"屋顶。

伯蒂奇诺大理石地砖等其他装饰材料的使用为设计增添了现代精致感和奢华感。

度假村共有193套客房，各式客房包括奢华客房、海滨俱乐部客房和少量的精选套房（配有巨大的私人露天平台）。

客房设计现代而经典，拥有宾客所需的最先进设施。灰褐、象牙白、鸭蛋蓝等柔和的色彩为房间营造出精致而放松的氛围。宽大的开放式布局大理石浴室和更衣区面向卧室开放，而露台则享有海洋和海岸暗礁所勾勒成的壮美景色。

1. Reception	11. Fitness centre	1. 接待处	11. 健身中心
2. Shops	12. Explorers club	2. 商店	12. 探险者俱乐部
3. Business centre	13. Reflections pool	3. 商务中心	13. 倒影池
4. Season	14. Reflections Restaurant	4. 四季餐厅	14. 倒影餐厅
5. Beach	15. Parking	5. 海滩	15. 停车场
6. Bar 68	16. Hotel entrance	6. 68酒吧	16. 酒店入口
7. Turtle bar	17. Mandara spa	7. 海龟酒吧	17. 曼达拉水疗中心
8. Main pool	18. Tennis courts	8. 主泳池	18. 网球场
9. Brezza bar	19. Water sports	9. 布勒沙酒吧	19. 水上运动中心
10. Brezza	20. Swimming zone	10. 布勒沙餐厅	20. 游泳区

1. Beach view 海滩景色	5. View of Reflections Restaurant from pool 从泳池区看回声餐厅景色
2. Resort plan 度假村平面图	6. Terrace on beach 海滨露台
3. Main lobby viewed from the beach 从海滩看大堂景色	7. Season Restaurant at night 四季餐厅夜晚景色
4. Brezza & Main pool view by night 布勒沙餐厅与主泳池夜晚景色	8. Season Restaurant 四季餐厅

9. Resort lobby at night
   度假村大堂夜景
10. The reception area
    接待区
11. Bar 68 entrance
    68酒吧入口
12. Brezza wine cellar private dining
    布勒沙酒窖私人用餐区

13. Spa entrance
水疗中心入口
14. Spa relaxation area outdoor
水疗中心户外休闲区
15. Spa lounge
水疗中心休闲吧

16、17. Spa treatment room
水疗中心治疗室
18. Couple guestroom
双人客房
19. Deluxe room ocean view
奢华海景客房
20. Spa suite
水疗套房

Long Beach Resort in Mauritius

毛里求斯长滩度假村

Completion date: 2011
Location: Belle Mare, Mauritius
Designer: Kevan Moses from Stauch Vorster
Architects (Architecture)
Keith Mehner from Keith Interior Design
(Interior Design)
Photographer: Long Beach Resort in Mauritius
Area: 238,765sqm

竣工时间：2011年
项目地点：毛里求斯，贝尔马尔
设计师：Stauch Vorster建筑事务所的Kevan Moses（建筑设计）
Keith室内设计公司的Keith Mehner（室内设计）
摄影师：毛里求斯长滩度假村
项目面积：238, 765平方米

Long beach is a major new resort opening on the east coast of Mauritius. Architecture brings tropical interpretations to contemporary urban themes, softened by natural forms, local detailing and materials. Buildings integrate with the resort's extensive lush tropical gardens. A unique architectural feature is the adjustable 'wind breaks' that modulate the offshore breezes. All elements of material work together, inside and outside, to give a unique feel to Long Beach.

The chic contemporary Long Beach designs blend together open-air and indoor living. This hotel brings

a totally new concept to the hotel environment in Mauritius and the region. A never-seen-before feel will immediately strike any guest with the 109sqm of beach space available for each room. The 255 rooms of Long Beach are arranged in three 'crescents' which are judiciously set so to enable maximum privacy. Every room has a view of the ocean and even the back rooms have been built on a raised level so that the stupendous view of the Belle Mare lagoon and the close-by ocean are available for every terrace. The rise of the full moon over this dazzling sea is a truly one of a kind experience.

The rooms are set to the tune of the hotel. Indeed, the space available in each room has been designed to enable guests to enjoy the best of their holidays. Magnificent king-size beds and stylish furniture harmonise well with the soft furnishings in tones of apple green, coral red, white and ash grey-white. All rooms have mini-bars, separate shower, toilet and bath with abundant space to breathe and move around. There are four restaurants on the Piazza, with outdoor and indoor seating, and another restaurant on the beach. Once more, the concept brought by the architects enables guests to enjoy an abundance

1. Main guest entrance	1. 客人主入口
2. Guest parking	2. 客人停车场
3. Guest arrival	3. 客人到达区
4. Lobby	4. 大厅
5. Reception	5. 接待处
6. Italian restaurant, Sapori	6. 萨波里意式餐厅
7. Main restaurant, Le March è	7. 马歇主餐厅
8. Lounge bar, Shores	8. 海岸高级酒吧
9. Japanese restaurant, Hasu	9. 波须日式餐厅
10. Chinese restaurant, Chopsticks	10. 中式筷子餐厅
11. Shops	11. 商店
12. Bombora Nightclub & Function room	12. 鲍姆保拉夜总会与功能间
13. Main pool (heated)	13. 主泳池（热水）
14. Fresh pool	14. 淡水泳池
15. Bar & Restaurant, Tides	15. 潮汐酒吧与餐厅
16. Spa	16. 水疗中心
17. Hammam	17. 哈曼浴室
18. Tennis Courts	18. 网球场
19. Multipurpose court	19. 多功能场地
20. Lap pool	20. 小型健身游泳池
21. Sports centre	21. 运动中心
22. Gym	22. 健身房
23. Teens club	23. 青少年俱乐部
24. Kids club	24. 儿童俱乐部
25. Boat house	25. 船坞
26. Beach	26. 海滩

of food experiences in cosy areas where bars also offer a lounge-like atmosphere to spend time alone, with loved ones or friends. The height of the ceiling at the Italian restaurant, the nooks of the main restaurant, the little zen-like 'forest' between the Piazza and the Chopsticks, the feet-in-the-sand experience of Tides: everything has been designed to make each outlet a unique destination. And the Piazza, this 'hub of activities' within the hotel is a very cosy place to be, as the central part of the whole hotel concept it will meet the desires of every guest.

Sea Spa is a tailor-made spa for Long Beach – salubrious, fit and intelligent. The Spa is an extension of the hotel and is set in an area which harboured a little endemic plants garden. The vegetation has been preserved, together with a pond, around which the spa buildings have been constructed, with a central hammam and separate treatment rooms. There are twelve treatment rooms including two double rooms and one equipped with bath. The outdoor spa pavilion is the ideal setting for massages overlooking the lagoon. Included is the Beauty Parlour – with hair salon, manicure and pedicure – a relaxation area and shopping.

Angels Kids Club with its 113sqm playroom is located in the Sports Centre and runs exciting and imaginative daily fun programmes with age-appropriate activities, facilities and amenities, twelve hours a day. The 148sqm Waves Teens Club is located in the Sports Centre area and offers free membership for 12-17 year olds with a special programme of sports and recreation as well as adventure outings and excursions.

1. Villa outdoor view
   别墅外景

2. Resort plan
   度假村平面图

3. Villa outdoor view
   别墅外景

4. Lobby entrance
   大堂入口

5、6. Villa outdoor view
   别墅外景

7. Bombora nightclub and function hall
   Bombora夜总会与功能厅

8. Resort main entrance
   度假村主入口

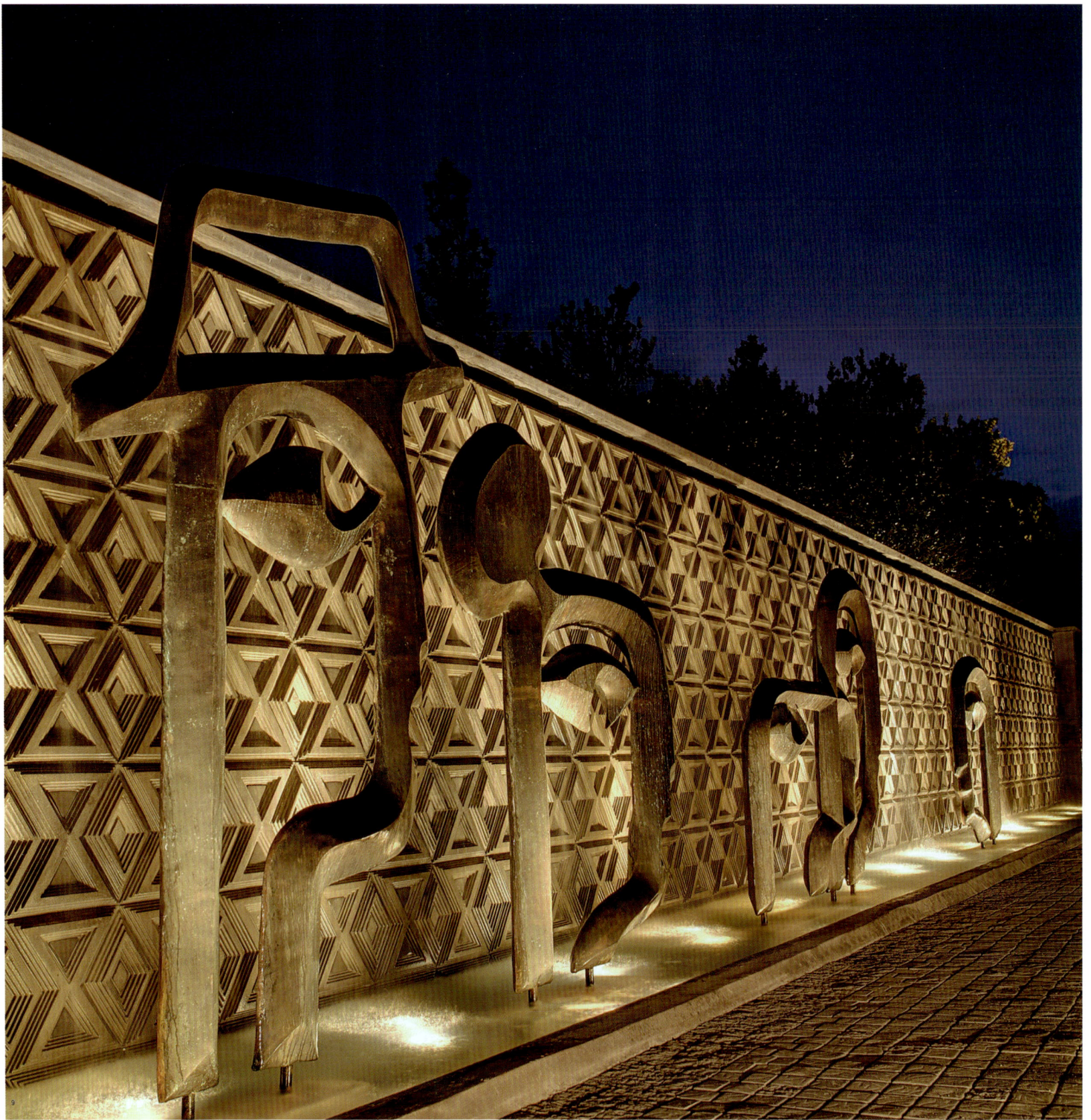

长滩度假村是毛里求斯东海岸新开业的一家大型度假村。建筑风格上是对现代都市主题在热带地区的重新诠释，又以自然的形式、细节设计和当地材料的选取使其变得更加柔和。所有的建筑物都与度假村广袤而苍翠繁茂的热带花园相结合。一个独特的建筑特色是可调节性"防风林"的设计，可以利用它来调节海风。室内外各种元素的材料相结合，彰显了长滩度假村的独特韵味。

长滩度假村别致而现代的设计手法使室内外的生活协调一致。该酒店给毛里求斯乃至整个该地区的酒店环境带来了一个全新的理念。客人们会被一种前所未有的感觉所冲击，他们会对每间客房可获得的109平方米的海滩空间感到震惊。长滩度假村内255间客房以三个新月型排列，这种精心的设计是为了创造最大的私密空间。每间客房都可以观赏海景，即使是位于后面的客房，建筑也会建在更高的地基上，这样就可以从每

个露台观赏到贝尔马尔环礁湖以及海滨附近的绝美景色。观赏圆月从耀眼的海洋上缓缓升起绝对是一次独一无二的体验。

客房的设计与整个酒店风格相协调。事实上，为了使客人能够享受最美好的假期，每间客房的可用空间都经过精心的设计。华丽的特大号双人床以及现代风格的家具与苹果绿、珊瑚红、白色及灰白色的软装饰风格协调统一。所有的客房都配有迷你酒吧，独立的淋浴间，卫生间及浴室，为客人提供充足的生活空间。

有四间餐厅建在露天广场上，提供室内外就餐空间，还有一间餐厅建在海滩上。建筑师的设计理念还是要使客人可以在舒适的地方享受丰富的美食体验，这里的酒吧也会提供一种休闲吧似的环境，让客人独自或与至亲好友共同度过休闲时光。意大利风格餐厅的天花板、主餐厅的各个角落、露天广场与Chopsticks中式餐厅之间的禅宗般的

小树林、Tides酒吧餐厅中脚踩沙滩的就餐体验，每间餐厅的设计都有其独具特色的地方。露天广场设计精妙，是酒店的活动中心。这是一个非常舒适的地方，是整个酒店设计理念的中心之处，可以满足每一位客人的需求。

海洋水疗中心是为长滩度假村特意量身打造的水疗中心——清爽、健康、智能。水疗中心是酒店的延伸，建在特有植物花园内一块隐秘之地上。保留着已有的植被和池塘，水疗中心的建筑围绕池塘而建，设有中心哈曼浴室和独立的治疗室。共有12间治疗室，其中包括两间双人治疗室，和一间带浴室的治疗室。户外水疗亭是按摩的理想之地，在这里可以俯瞰坏礁湖。还有美容室——美发沙龙与美甲室 是放松休闲与购物的理想之地。

天使儿童俱乐部位于运动中心，有113平方米的游戏室，设有12小时开放的、适合儿童

的、充满刺激与想象的娱乐活动项目与设施。148平方米的海波青少年俱乐部位于运动中心，为12到17岁的免费会员提供特别的运动与娱乐项目和冒险与远足活动。

9. Wall feature at main entrance
   主入口的墙上特色
10. Landscape
    景观
11. Wall feature
    墙上特色
12. Set up at fresh pool
    淡水泳池休息区
13. Kids club
    儿童俱乐部

14. Spa entrance
    水疗中心入口

16. Spa massage room
    水疗中心按摩室

15. Sea spa room
    深海水疗室

17. Sea spa terrace
    深海水疗露台

18. Shores bar
海滨酒吧
19. Sapori Restaurant
萨波里餐厅
20. Le marche buffet
马尔什自助餐厅
21. Table set up outside Sapori Restaurant
萨波里餐厅厂外餐桌
22. Guestroom outside
客房外景

23. Guestroom outside
    客房外景
24. View from bedroom
    从卧室看外景
25. Guest room
    客房

26. Couple guest room
    双人间客房
27. Living room
    客厅
28~30. Guest room
    客房

Sugar Beach Resort in Mauritius

毛里求斯蜜糖海滩度假村

Completion date: 2008(renovation)
Location: Port Louis, Mauritius
Designer: Macbeth Architects (Architecture)
Wilson & Associates (Interior Design)
Photographer: Sugar Beach Resort in Mauritius
Area: 120, 000sqm

竣工时间：2008年（翻新）
项目地点：毛里求斯，路易斯港
设计师：Macbeth建筑事务所(建筑设计)
Wilson联合设计公司（室内设计）
摄影师：毛里求斯蜜糖海滩度假村
项目面积：120, 000平方米

Sugar Beach is on the Leeward West coast of Mauritius. With over half a kilometre of sandy beaches, the resort covers more than 120, 000 square metres featuring beautiful landscaped tropical gardens with contemporary plantation-style architecture. The buildings comprise the central reception, the Manor House and 16 villas — 10 with 12 rooms each and 6 with 10 rooms each.

Sugar Beach is the perfect place to relax, with 30 acres of beautiful gardens, all manner of water and land sports and activities by day, the romance of the starlit tropical sky at night and glorious food with a feast of

Mauritian, Mediterranean and world flavours. The range of facilities makes it ideal for incentives, as well as couples and families to experience the best of famed Mauritian hospitality.

There are 258 guest rooms and suites. Interior design complements the plantation theme of the resort, while maintaining a contemporary feel. Real mahogany furniture and magnificent king-size beds harmonise well with the soft finishes in tones of pale green and beige. The rooms are fitted with blackout curtains and pastel sheers drawn behind white louvered panels on each side of the front windows. The sheers maintain privacy while allowing filtered light to enter during the day.

Standard rooms have bath-with-shower and toilet. Superior rooms have separate bath, stand-up rain shower cubicle, separate toilet and a larger vanity. All rooms have their own private patio or balcony complete with easy chair, ottoman and day bed. All rooms are provided with fully stocked mini-bars, luggage racks, tea and coffee making facilities and LCD fat-screen TVs.

The Spa comprises both single and couple treatment rooms, and also has a dedicated Shiatsu treatment

area. For the ultimate therapy, try the deluxe hammam treatments including an exotic black soap massage. Visitors to the Spa may relax around a plunge pool and enjoy refreshments from the Organic Spa Bar. A fully equipped hairdressing salon for men and women is also located in the Spa area.

1. Reception area
2. Mon Plaisir Restaurant
3. Mon Plaisir Bar
4. The Emporium
5. Mon Plaisir Court Yard
6. Manor House lobby & lounge
7. Spa, hairdresser and hammam
8. Fitness centre
9. Sports bar
10. Citronella's café
11. Boat House
12. Kids club
13. Convention centre
14. Tides Beach Restaurant & Bar
15. North villas
16. South villas

1. 接待区
2. 乐宫餐厅
3. 乐宫酒吧
4. 商场
5. 乐宫庭院
6. 庄园主住宅大厅与休闲吧
7. 水疗中心、美发中心与哈曼浴室
8. 健身中心
9. 运动酒吧
10. 香茅咖啡厅
11. 船坞
12. 儿童俱乐部
13. 会议中心
14. 潮汐海滩餐厅与酒吧
15. 北部别墅区
16. 南部别墅区

蜜糖海滩度假村位于毛里求斯西海岸下风方向。该度假村覆盖面积达120000平方米，有半公里多长的沙滩、美丽的热带花园景观和现代种植园风格的建筑。建筑群由中央接待处、庄园住宅和16套别墅组成——其中10套别墅有12间客房，另外6套别墅有10间客房。

蜜糖海滩度假村是放松休闲的完美之地，有30英亩的花园景观，白天有各种各样的水上和陆地上的体育运动和活动，晚上有热带天空上星光闪耀的浪漫景色，还有毛里求斯、地中海及世界各地的特色风味美食。一系列的服务设施使这里成为激励自己的理想之地，也是夫妇或全家人体验毛里求斯闻名的好客之道的最佳场所。

度假村共有258间客房和套房。室内设计保持现代风格，同时紧扣该度假村种植园的主题。真正的红木家具和华丽的超大号床与淡绿和米黄色调的软装饰和谐统一。客房装有遮光窗帘，前窗两侧白色百叶窗板的后面是柔和的透明织物。这种透明的窗帘既可以在白天让一部分阳光照射进来，又可以保持一定的私密性。

标准客房配有带淋浴的浴室和卫生间。高级客房有独立的浴室、淋浴间、独立卫生间和更大的台面。所有的客房都有自己的露台或阳台，上面摆放着安乐椅、搁脚凳和沙发床。所有的客房都配有功能俱全的迷你酒吧、行李架、茶和咖啡调制设施和LCD宽屏电视。

水疗中心包含单人治疗室和双人治疗室，还有一个指压按摩专用区。最根本的治疗方式是高级哈曼治疗，包括一种异国情调的黑肥皂按摩。该水疗中心的游客可以在跌水潭周围休息放松，享受来自有机水疗吧的点心。设备齐全的美发沙龙也位于水疗中心。

1. Resort pool and landscape
   度假村泳池与景观
2. Resort plan
   度假村平面图
3、4. Pool view
   泳池景观
5. Pool view and deck
   泳池与甲板区
6. General view of the resort
   度假村总貌
7. Pool at night
   泳池夜景
8. Pool Aura Spa outside
   户外水疗池
9. Spa pool inside
   室内水疗池

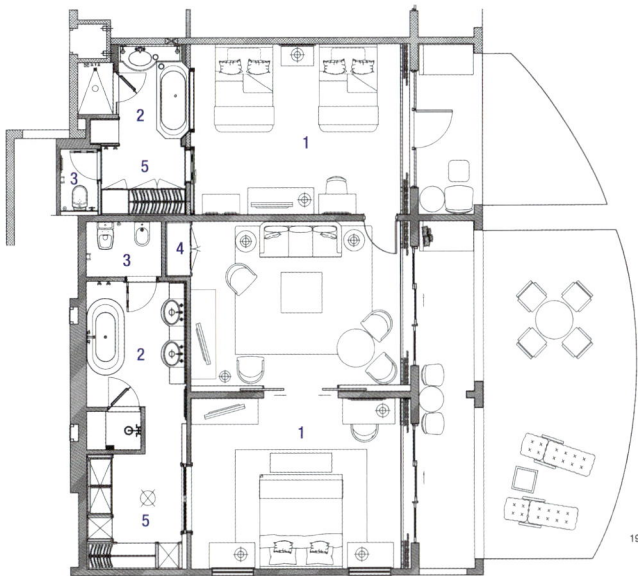

1. Bedroom
2. Bathroom
3. WC
4. Bar closet
5. Wardrobe

1. 卧室
2. 浴室
3. 卫生间
4. 酒吧壁橱
5. 衣橱

10、11. Lobby lounge
　　　 大堂吧
12. Citronella Restaurant
　　 Citronella餐厅
13. Tides Restaurant
　　 潮汐餐厅
14. Lobby lounge
　　 大堂吧
15. Villa exterior
　　 别墅外观
16. Living room
　　 客厅
17、18. Guest room
　　　 客房
19. Villa plan
　　 别墅平面图

Le Touessrok Resort in Mauritius

毛里求斯迪拉突斯洛克度假村

Completion date: 2007(renovation)
Location: Port Louis, Mauritius
Designer: Archos Tekten Management Co Ltd (Architects),
Hirsch Bedner Associates (Villas)
Bernhard Langer (Golf Course)
Photographer: Le Touessrok Resort in Mauritius
Area: 250, 000sqm

竣工时间：2007年（翻新）
项目地点：毛里求斯，路易斯港
设计师：Archos Tekten管理有限公司（建筑设计）
Hirsch Bedner联合设计公司（别墅设计）
Bernhard Langer（高尔夫球场设计）
摄影师：毛里求斯迪拉突斯洛克度假村
项目面积：250, 000平方米

Le Touessrok is one of the world's greatest resorts — at the height of cool, modern elegance, imbued with the warmth of tropical Mauritius, it is truly one of the 'Leading Hotels of the World'. It lies on a beautiful stretch of sandy coast, looking over the tranquil Trou d'Eau Douce Bay. Out in the lagoon are two beautiful islands, including Ile aux Cerfs with its spectacular 18-hole championship golf course.

Regulars and new guests who fly in from across the world to discover this mythical hotel are greeted by a dramatic, light filled lobby with views straight out across the free-form swimming pool to the lovely

curve of the bay. Walls of off-white stucco are set off by chocolate brown wicker furniture and splashes of distinctive colour. An elegant wooden bridge crosses to Frangipani Island, with its fabulous Givenchy Spa and Fitness Centre. Barlen's — the ultra-casual beachside restaurant — looks directly over the bay. All the resort's 132 suites and 68 rooms have an unimpeded sea view.

The suites are airy split-level layouts with chic off-white walls and porcelain-tiled floors, crisp white Egyptian cottons on the beds, bathrooms with funky egg shaped baths of poured marble (with sea views),

fibre optic lighting, contrasting with rich African woods, soft fabrics and decorative panels reminiscent of traditional Creole designs.

There are three spectacular new waterfront villas which redefined the benchmark for luxury in Mauritius. The designer trio was conceived to satisfy the current burgeoning demand for ever more exclusive and expansive accommodation that combines high technologies with high expectations. Each villa has three spacious bedrooms and en-suite bathrooms, making them the perfect retreat for families, couples or friends. Le Touessrok has

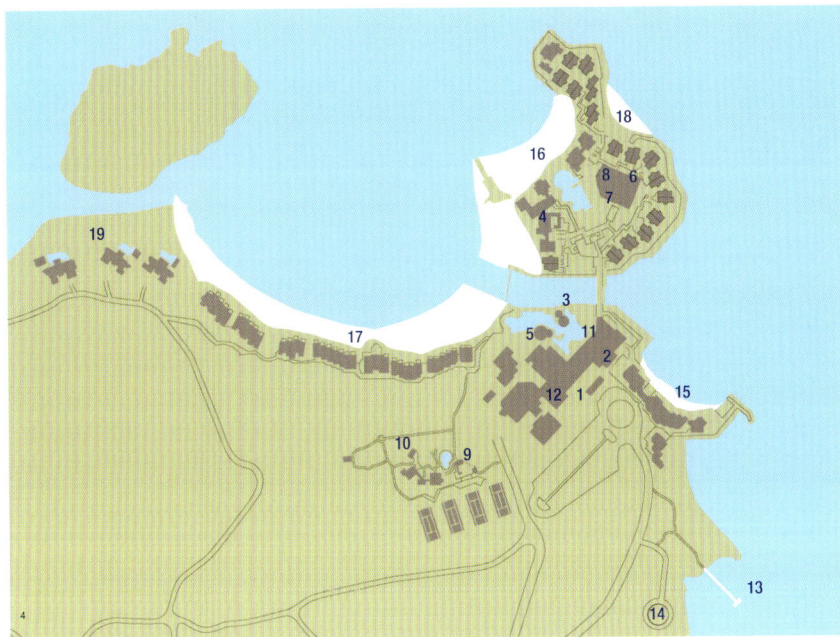

1. Main entrance
2. Three-Nine-Eight
3. Safran
4. Barlen's
5. Sega Bar
6. Givenchy Spa
7. Hairdressing Salon
8. Fitness centre
9. Peter Burwash International Tennis and Recreation Centre
10. T-Club
11. Diving centre
12. Shopping arcade
13. Jetty to Ile aux Cerfs, Golf Course and Ilot Mangenie
14. Helipad
15. Coral beach
16. Barlen's beach
17. Hibiscus beach
18. Frangipani beach
19. The villas

1. 主入口
2. 三九八餐厅
3. 撒弗朗餐厅
4. 巴伦斯餐厅
5. 世嘉酒吧
6. 纪梵希水疗中心
7. 美发沙龙
8. 健身中心
9. 彼得博尔西国际网球与娱乐中心
10. T俱乐部
11. 跳水中心
12. 购物商场
13. 通往瑟芬小岛、高尔夫球场与曼阁涅小岛的码头
14. 直升机场
15. 珊瑚礁海滩
16. 巴伦斯海滩
17. 芙蓉花海滩
18. 赤素馨花海滩
19. 别墅区

embraced the residential-style villa concept and given it an extra special twist.

All the villas have been designed to mirror the resort's signature style. Yet each can cite its own individuality with distinctive finishes: a seductive swimming pool, a garden filled with flowers and lush foliage, direct access to the beach, a secluded terrace and a separate entrance. Contemporary interpretations of Mauritian ethnic chic combine with timbered decks, thatched roofs, whitewashed walls and ceramic floors.

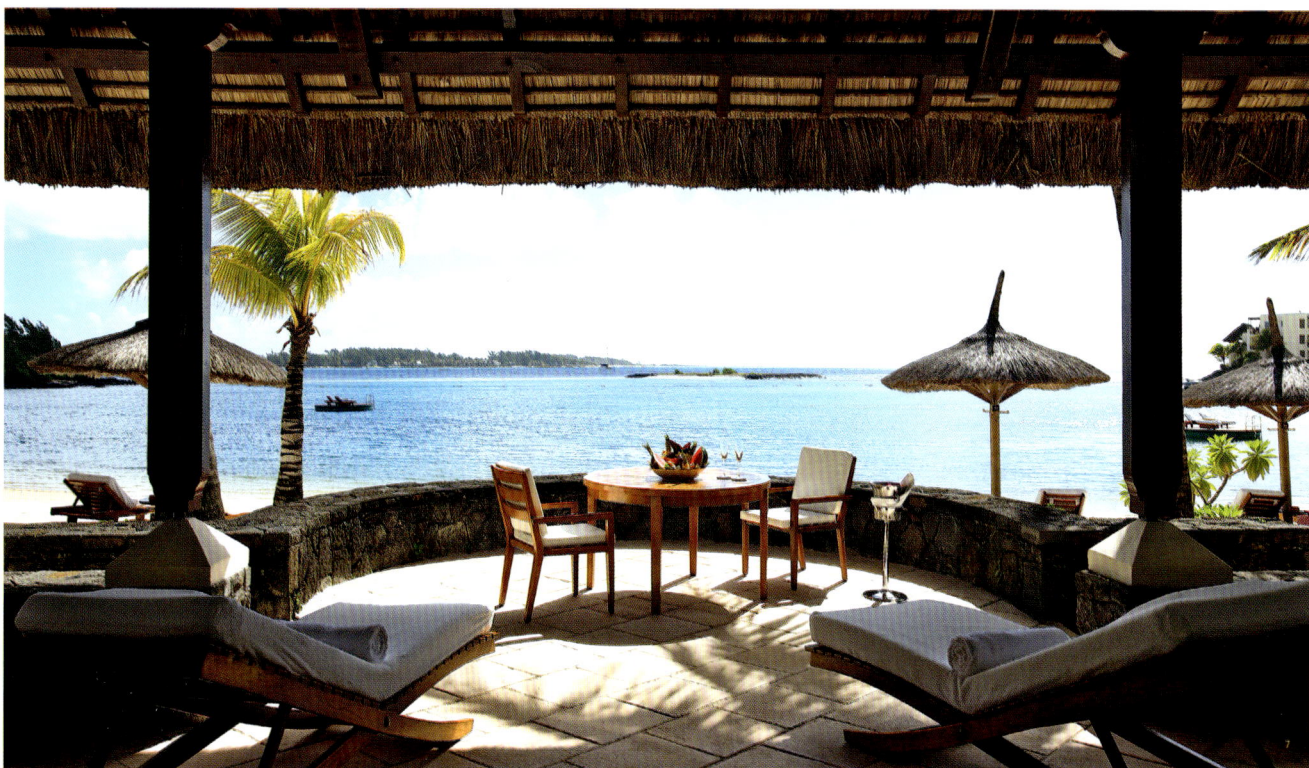

迪拉突斯洛克度假村是世界上大型度假村之一，清新现代的典雅中渗透着毛里求斯的热带温情，是真正的世界顶级酒店。该度假村位于美丽绵长的沙质海岸上，可以俯瞰迪鸥岛斯海湾。环礁湖内有两个美丽的小岛，其中瑟芬小岛上建有18洞高尔夫球场。

来自世界各地的常客和新游客们到达这个充满神秘的酒店，首先来到引人注目的灯光闪耀的大堂，在这里可以直接眺望到无规则泳池和海湾柔美的曲线。灰白色的墙面以巧克力棕色柳条家具和喷溅的独特色彩为衬托。典雅的木桥通往赤素馨花岛上的纪梵希水疗中心和健身中心。巴伦斯休闲海滨餐厅可以直接观赏海湾景色。度假村中的132间套房和68间客房都可以直接观赏到海景。

套房内的通风布局错落有致，米白色的墙面和瓷砖地面，纯白色的埃及棉床品，装有蛋形大理石浴缸的浴室（可以观赏海景），光纤照明设备，与昂贵的非洲木，柔软的面料和克里奥尔传统的装饰板设计形成鲜明的对比。

度假村有三套新建的海滨别墅，华丽而壮观，重新定义了毛里求斯地区奢华的基准。设计师的设计构想是满足目前迅速增长的对专属而宽阔的高科技、高水准客房的需求。每套别墅有三间宽敞的卧室，套房浴室，是家人、夫妇或朋友们完美的避世之地。独木桥度假村别墅的设计理念是在住宅式别墅基础上做了一些特别的改变。

所有的别墅都反映了度假村的本色风格，但每套别墅又都用独特的装饰来彰显自己的个性特征·齐满诱惑力的游泳池，种满鲜花与植物的花园，直接通往海滩的小路，隐蔽的露台和独立的入口。木制露台，茅草屋顶，白色的墙壁和陶瓷地面，所有这些都是对毛里求斯别致风格的现代诠释。

8. Villa terrace beach view
   别墅露台海景
9. Barlens Restaurant
   巴伦斯餐厅
10. Golf club
   高尔夫俱乐部
11. Barlens Bar
   巴伦斯酒吧
12. Barlens Restaurant
   巴伦斯餐厅

13. The spa entrance
水疗中心入口
14. Massage pavilion outdoor
户外按摩亭
15. Spa relaxation area
水疗中心放松区
16. Spa reception
水疗中心接待处
17. Spa pool detail
水疗池细部

18. Villa terrace
    别墅露台
19. Lounge
    休闲吧
20. Villa living area
    别墅客厅
21. Villa bedroom
    别墅卧室
22. Villa bathroom
    别墅浴室
23. Villa plan
    别墅平面图

1. Living room
2. Single bedroom
3. Double bedroom
4. Bathroom
5. Terrace

1. 起居室
2. 单人间卧室
3. 双人间卧室
4. 浴室
5. 阳台

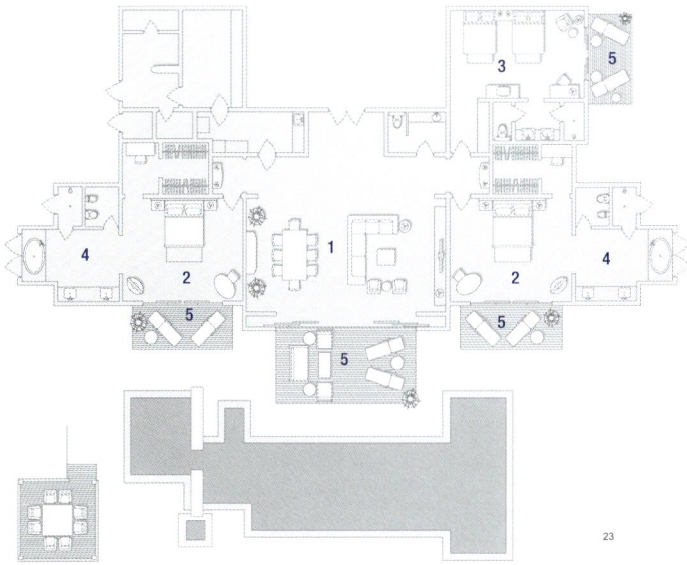

La Pirogue Resort in Mauritius

毛里求斯独木舟度假村

Completion date: 2003 (Redesigned)
Location: Flic en Flac, Mauritius
Designer: Alister Macbeth Architects
Photographer: La Pirogue Resort in Mauritius
Area: 140,000sqm

竣工时间：2003年（重新设计）
项目地点：毛里求斯，福莱
设计师：Alister Macbeth建筑事务所
摄影师：毛里求斯独木舟度假村
项目面积：140,000平方米

Located on the west coast of Mauritius, La Pirogue is set in the midst of a vast, luxuriant tropical garden. Close to Flic-en-Flac, the architecture has been inspired by the sea and features a fishing village atmosphere, starting with the fishing boat, from which the hotel gets its name, to the main building inspired by white sails floating in the wind and to the chalets built in local style with volcanic rock and decked with wood and thatch.

Its main building continues the theme, with a half-dome facing out across the pool to the sea like a ship. The guest cottages, set in discreet clusters deep in

the gardens or facing the beach, are also shaped like sailing boats. The lobby is calm and cool, leading through to the restaurants, pool and the sea.

The island looking thatched cottages, each with a private terrace are scattered about the beautiful gardens, in crescents and horseshoes. Guests love the fact that every cottage is ground-level, giving a private feel and making it convenient for families. The colour scheme looks to the sea and its shells for inspiration, combining creamy vanilla and coral shades on wide-striped bedspreads and lampshades. Sweeping curtains with coral edging are set off by cool, off-white walls and ceilings – making the rooms as bright as possible. Standard rooms are two to a cottage, with cool floors, airy ceilings, bathrooms with separate showers, and floor to ceiling French windows and balconies, separated from the adjoining room by a trellis and lush plants. Most Superior Rooms occupy an entire cottage, with larger bathrooms and an exclusive terrace. The Royal Suites, named Flamboyant and Bougainvillea, boast large living and dining areas, double terraces and spacious double rooms and bathrooms with Jacuzzi, guest toilet, and a discreet position set back in the gardens for additional privacy.

1. Thatches Restaurant
2. Restaurant La Badiane
3. Blue Bar
4. Entertainment Area
5. Shops
6. Towel Cabana
7. Coco Kiosk Bar
8. Pool
9. Beach Bar
10. Paul & Virginie Restaurant
11. Sun Kids Club
12. Diving centre
13. Recreational centre
14. Big Game Fishing Desk
15. Boat House
16. Mini golf area

17. Petanque/Badminton/Basketball
18. Citronella's Café
19. The Spa, Hairdresser and Hammam
20. Tennis club, Sports bar
21. Fitness centre
22. Reception & Information desk and Post office box
23. Conference room—Muscade
24. Board room—Cardamome
25. Clinic
26. Conference room—Sesame
27. Tour operators desks
28. Parking
29. Entrance
30. Wedding gazebo
31. Poney club

32. Football Pitch
33. 'Le Bougainvillaea' Royal Suite
34. 'Le Flamboyant' Royal Suite
35. Taxi
36. Archery
37. North wing rooms
38. South wing rooms

1. 茅草屋顶餐厅
2. 八角茴香餐厅
3. 蓝调酒吧
4. 休闲娱乐区
5. 商店
6. 有凉台的小屋
7. 椰树凉亭酒吧
8. 游泳池

9. 海滨酒吧
10. 保罗与弗吉尼亚餐厅
11. 阳光儿童俱乐部
12. 潜水中心
13. 娱乐中心
14. 大型钓鱼比赛区
15. 船坞
16. 小型高尔夫场
17. 法式滚球场、羽毛球场、篮球场
18. 香茅咖啡厅
19. 水疗中心、美发中心和哈曼浴室
20. 网球俱乐部与运动酒吧
21. 健身中心
22. 接待处与信息咨询台和邮政信箱
23. 马斯凯德会议室
24. 卡德毛姆董事会会议室

25. 诊所
26. 赛瑟米会议室
27. 旅行社接待处
28. 停车场
29. 入口
30. 婚礼亭
31. 波尼俱乐部
32. 足球场
33. 九重葛皇家套房
34. 凤凰皇家套房
35. 出租车场地
36. 射箭场地
37. 北翼客房
38. 南翼客房

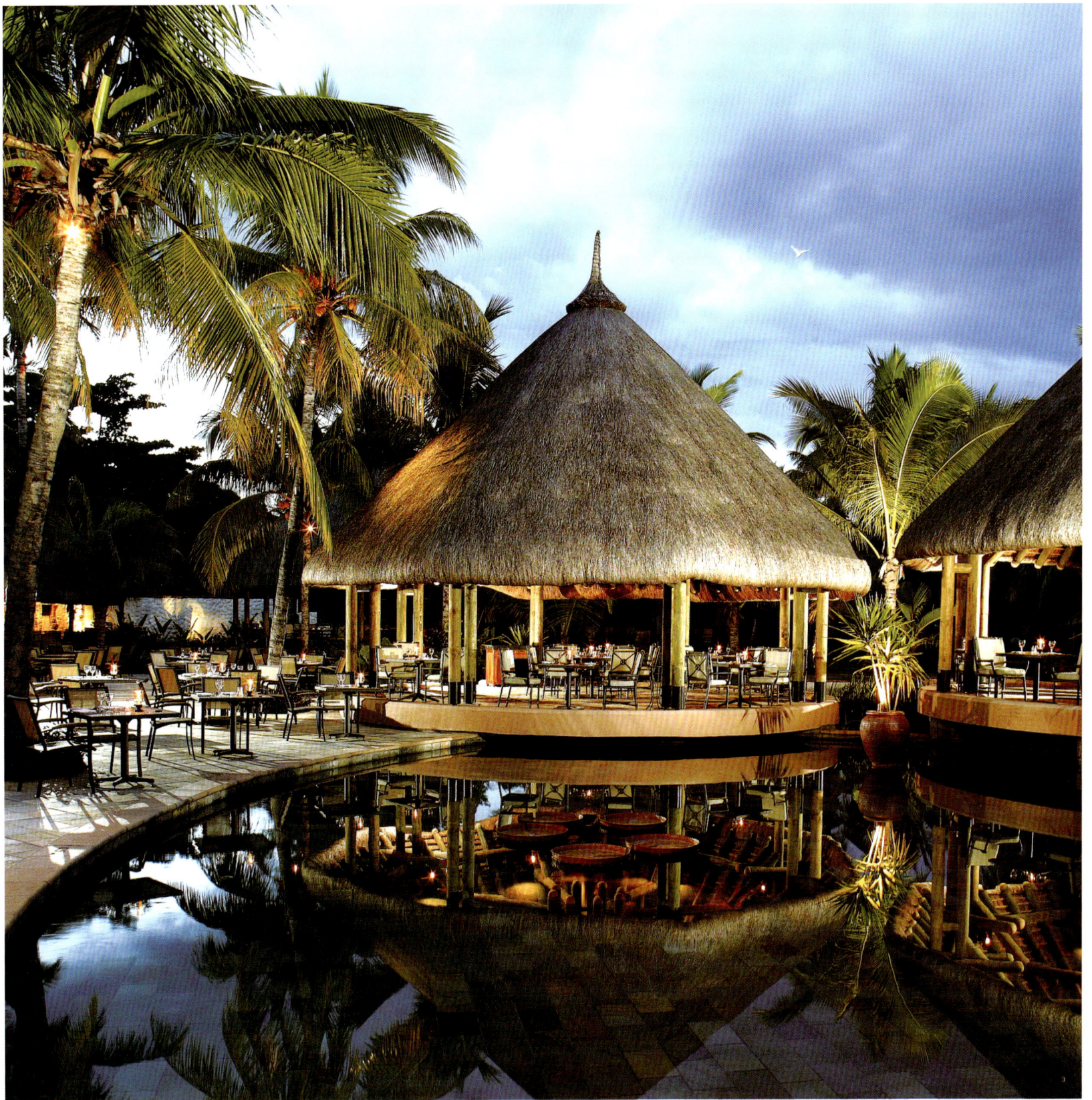

The decor is minimal, as befits a tropical beach hotel, combining the soaring roofs and bright light from the sea and sand outside with cool, airy colours, painted to look gently faded in the sun.

There are three restaurants, three bars, some conference facilities, a Kids & Teens Club and a very popular programme of nightly entertainment performed by the hotel's talented show team. The Spa comprises both single and couple treatment rooms and also has a dedicated Shiatsu treatment area. For the ultimate therapy try the deluxe Hammam treatments including an exotic black soap massage. A fully equipped hairdressing salon for men and women is also available in the Spa area.

独木舟度假村位于毛里求斯西海岸，掩映在一片广袤繁盛的热带花园中。酒店离福莱很近，建筑风格受海洋的启发，形成一种渔村环境。酒店就是以渔船的名字来命名的，主建筑的设计灵感来自风中扬起的白帆，居住的小屋是以当地的建筑风格、用火山岩建成的，并以木材和茅草作为装饰。

度假村的主建筑以此为主题，半圆的屋顶朝向泳池和大海，犹如一艘船在海中航行。客房小屋或集群地建于花园深处，或面海而建，犹如海中航行的帆船。大堂宁静而清爽，可以通往餐厅、游泳池和海边。

每间茅草屋顶的小屋都有一个私人阳台，这些小岛一样的小屋以新月形或马蹄形建于美丽的花园中。每间小屋客房都位于一层，给人一种私密感，更方便全家人居住，广受客人的喜爱。色彩设计的灵感来自海洋和海中的贝壳，奶油香草色和珊瑚色搭配使用在宽条床罩和灯罩上。带有珊瑚边饰的落地窗帘，与清爽的米白色墙面和天花板相衬托，使房间显得越发明亮。

一个小屋有两间标准客房，有清爽的地板，通风的天花板，带独立淋浴间的浴室，法式落地窗和由一个特色花架和繁茂的植物隔开的两间客房阳台。大部分的高级客房都独占整个小屋，配有更大的浴室和独享的阳台。名为Flamboyant和Bougainvillea的两间皇家套房拥有宽敞的起居室和餐厅，双阳台和宽敞的双人间卧室，和带极可意水流按摩浴缸的浴室，客用卫生间，还有在花园中另外设置的私密场所。装饰风格适合热带海滨酒店，高耸的屋顶和来自海洋与沙滩那种清爽而明快的色彩相结合，这种油漆的色彩在阳光下看起来有些褪色。

度假村有三间餐厅，三间酒吧，各种会议设施，一个儿童及青少年俱乐部和一个非常受欢迎的夜间娱乐节目，这个节目是由酒店精英表演团队表演的。水疗中心包含单人治疗室和双人治疗室，还有一个指压按摩专用区。最根本的治疗方式是高级哈曼治疗，包括一种异国情调的黑肥皂按摩。设备齐全的美发沙龙也位于水疗中心。

1. Resort aerial view
   度假村鸟瞰图
2. Resort plan
   度假村平面图
3. Poolside restaurant
   泳池边餐厅
4. Restaurant pavilion
   餐饮亭
5. Beach bar
   海滨酒吧
6. Beach restaurant
   海滨餐厅
7. Royal suite exterior
   皇家套房外观
8. Sun Kids club
   儿童俱乐部
9. Wedding pavilion
   婚礼亭
10. Superior guest room
   高级客房

11. Lobby by night
    夜间大堂
12. Cafeteria
    自助餐厅
13. Suite living room
    套房客厅
14. Standard guest room
    标准客房
15. Suite bedroom
    套房卧室

Sheraton Gambia Hotel Resort & Spa

喜来登冈比亚水疗度假酒店

Completion date: 2007(Renovation)
Location: Banjul, Gambia
Designer: Quadtrisir Architects
Photographer: Peter Jackson
Area: 26, 220sqm

竣工时间：2007年（翻新）
项目地点：冈比亚，班珠尔
设计师：Quadtrisir建筑事务所
摄影师：彼得·杰克逊
项目面积：26, 220平方米

Sheraton Gambia Hotel Resort & Spa is a spectacularly beautiful resort inspired by local traditions on an unspoilt beach on the Atlantic Ocean. Built to resemble a typical African village, immersed in lush garden of palm and baobab trees, the Sheraton Gambia Hotel Resort & Spa brings you the authentic spirit of West Africa.

The lower level of the hotel is situated next to the unspoilt beach with lawns surrounding the large infinity swimming pool. There are many pathways and steps within the hotel, and two glass lifts connecting the reception, Cremetata buffet restaurant

and the pool/bar area. The Sheraton Hotel is built into the hillside, nestled between giant baobab trees. Its rooms are of contemporary style, mostly enjoying sea views from their ocean-facing balcony or terrace. There are 2 wings: the Baobab Wing is closer to the main building and restaurants housing the Standard Baobab rooms, which are fairly compact. There is also a variety of upgraded room options which are larger and closer to the beach. The tranquil, more exclusive Palm Wing houses the cliff-top Sunset Palm rooms as well as the superbly located, beachfront Beach Palm rooms, both of which are larger with a lounge area and offer uninterrupted ocean views from both ground and upper floors.

The Sheraton Gambia Deluxe Room captures the very essence of luxury, timeless grace and sophistication. A perfect setting that is both intimate and relaxed and conducive to business and leisure. Furnished in contemporary African design, each deluxe room offers unmasked sea views right from the comfort of the Sheraton Sweet Sleeper bed or opens to the refreshing breeze of the crystal blue Atlantic ocean.

The luxurious Suites have separate bedroom & sitting room both with magnificent ocean & sunset views.

1. Presidential Suite
2. Sunset Palm rooms and suites
3. Beach Palm rooms and suites
4. Classic rooms
5. Sunset Baobab rooms
6. Beach Baobab rooms
7. Treetop lounge bar
8. Crematata Restaurant (1st floor)
9. Bar A Kuda
10. Balaton Restaurant
11. Swimming pool
12. Children's pool

1. 总统套房
2. 落日棕榈客房与套房
3. 海滨棕榈客房与套房
4. 经典客房
5. 落日猴面包树客房
6. 海滨猴面包树客房
7. 树顶休闲酒吧
8. 克莱姆塔塔餐厅（二层）
9. 阿酷达酒吧
10. 巴拉顿餐厅
11. 游泳池
12. 儿童游泳池

The Suites are designed with the guest's comfort and needs in mind with state-of-the-art facilities. The beautiful African décor gives the room a light and spacious atmosphere. The Presidential Villa offers pure luxurious space and incredible views of the sea from the cliff top. It consists of two spacious bedrooms, a dining room, a living room, a TV area, a work area and a guest toilet. This is the perfect disposition for any guest with exclusive needs and is also ideal for hosting business lunches or important dinner engagements.

The Spa is located in a separate building with indoor and outdoor treatment areas, from where you can enjoy a massage looking down at the sea. There is also a quiet pool in this area overlooking the beach. The spa offers a good choice of treatments and especially includes a massage overlooking the beach in one of the outdoor treatment cabanas.

1. Terrace with sea view
   海景露台
2. Resort plan
   度假村平面图
3. Massage area of the spa
   水疗中心按摩区
4. Spa pool
   水疗池
5. Spa relaxing room
   水疗中心休息室
6. Crematata Restaurant
   Crematata餐厅
7、8. Lobby bar
   大堂吧
9、10. Presidential Villa
   总统别墅

喜来登冈比亚水疗度假酒店位于大西洋海岸天然的海滩上，拥有壮观秀丽的景色，设计灵感来自当地的传统。仿照典型的非洲村落而建，掩映于花园内郁郁葱葱的棕榈树与猴面包树间，喜来登冈比亚水疗度假酒店带你体验真正的西非风情。

酒店较低层的设施临天然海滩而建，有一个大的无边泳池，四周草坪环绕。酒店内有很多小路与台阶，还有两部玻璃电梯，通往接待处、Cremetata自助餐厅和泳池区或酒吧区。喜来登酒店建在山坡上，掩映在巨大的猴面包树间。客房采用的现代的设计风格，从面海的阳台或露台可以享受美丽的海景。客房分为两翼：猴面包树翼靠近主建筑和餐厅，这里有相当简洁的猴面包树标准客房。还有各种升级客房，空间更大，更靠近海滩。棕榈树翼是宁静而更加专属的客房区，有悬崖顶部的落日棕榈客房和华丽的海滨棕榈客房，这两种客房都有休闲区，空间更大，从一层到上面的楼层都可以欣赏一览无余的美丽海景。

喜来登冈比亚豪华客房设计秉承了奢华、永恒的优雅与复杂的宗旨，这里亲密而又放

松的完美环境有益于业务往来和休闲放松。每间豪华客房都以现代非洲设计风格的家具装饰，从舒适的喜来登甜梦睡床上可以欣赏美丽的海景，享受来自湛蓝的大西洋清新的海风。

奢华的套房有独立的卧室和起居室，可以欣赏壮观的海景和日落时分的美丽景色。套房配有最顶级的设备，一切设计都是为了满足客人的舒适与需求。漂亮的非洲装饰使客房显得更加明亮和宽敞。总统别墅提供了纯奢华的空间，并可以从悬崖顶眺望惊人的海景。总统别墅包含两间宽敞的卧室、餐厅、起居室、电视区、工作区和客用卫生间。这种完美的格局设置是为那些有专属需求的客人准备的，这里是举办商务午餐或重要的晚餐约会的理想之地。

水疗中心位于一个独立的建筑中，有室内外治疗区，可以在享受按摩的同时俯瞰大海。还有一个宁静的泳池区，在这里可以俯瞰海滩。水疗中心提供各种水疗方案，特别推荐的是在户外治疗区俯瞰海滩的按摩。

La Sultana Marrakech

马拉喀什苏丹娜酒店

Completion date: 2004
Location: Marrakech, Morocco
Designer: La Sultana Concept
Photographer: La Sultana Hotels
Area: 4, 700sqm

竣工时间：2004年
项目地点：摩洛哥，马拉喀什
设计师：苏丹娜概念设计公司
摄影师：苏丹娜酒店集团
项目面积：4700平方米

In the heart of imperial Marrakech, in the first kasbah of the sultans, next to the ancient Saadian tombs, is the hotel La Sultana. It is an oasis of calm and comfort amidst the swirl and bustle of the Medina. The garden of the Scheherezade has the lushness of the African savannah. The salons and suites are decorated in Senegalese style. La Sultana shows Moroccan craftsmanship at its best, from the intricate stucco ornamentation to the hand-carved woodwork and tadelkat (polished plaster) surfaces. Every part of the hotel has its own associations and atmosphere. When you open the door to your bedroom you enter

a world of fantasy. The 28 bedrooms and suites all have their own individual style but they share the La Sultana hallmark of oriental splendour. The suite, with its bold fuchsia drapes and carpets, its slate-blue tadelakat and carved cedarwood furniture, combines the worlds of Indian maharajah and Moorish sultan. Rooms and suites feature their own decoration mingling sophistication with detailed perfection and elegance of an intimate atmosphere. There are marble or Zellige bathrooms with bath and shower and separate toilet.

The spa and beauty parlour offer every type of treatment for health and relaxation. As well as an exercise room, sauna and Jacuzzi there are shared and individual hammams, massage rooms and lounges. There is 1 individual hammam, 1 hammam for couples and 1 hammam for 4 persons, relaxation area around Jacuzzi, Beauty space, 2 beauty cabins, 5 massage rooms, 2 of them outdoor, 1 Balneotherapy room, Sauna Finlandais, professional team with 8 therapists.

苏丹娜酒店位于马拉喀什市中心第一个苏丹原住居民区内，紧邻Saadian古墓群，是麦地那喧嚣的漩涡中一片宁静而舒适的绿洲。Scheherezade花园有着非洲大草原一样的繁茂，各沙龙和套房采用的是塞内加尔风格的装饰。苏丹娜酒店是摩洛哥手工艺最好的展示，从错综复杂的灰泥装饰，到手工雕刻的木制品和表面抛光的石膏工艺品，酒店内每个部分都有其独特的关联和氛围。

当你开门进入客房，犹如走进一个梦幻的世界。28间客房和套房都有各自独特的风格，但同时又都具有苏丹娜酒店东方华美的特点。套房结合了印度大君式风格和摩尔苏丹式风格，装饰有樱红色的窗帘和地毯，暗蓝色的抛光石膏工艺品和雕刻杉木家具。客房和套房有各自独特的装饰风格，将复杂的装饰与完美而典雅的细部设计相融合，营造亲密的氛围。客房设有带浴缸和淋浴设施的大理石浴室和独立的卫生间。

水疗与美容中心提供各种类型的健身与放松治疗，除了一间运动室、桑拿浴室和极可意水流按摩浴缸以外，还有共享与独立的哈曼浴室、按摩室与休闲吧。其中有一间单人哈曼浴室，一间双人哈曼浴室和一间4人间哈曼浴室，还有极可意水流按摩浴缸周围的休闲区，美容区，两间美容小屋，5间按摩室（其中两间为室外按摩室），一间治疗室，芬兰桑拿浴室以及8位治疗专家组成的专业团队。

1. Swimming pool
游泳池
2. Resort plan
度假村平面图
3. Atrium
中庭

1. Reception	1. 接待处
2. Boutique	2. 精品店
3. Salon	3. 沙龙
4. Spa	4. 水疗中心
5. Jaccuzi	5. 极可意水流按摩浴缸
6. WC	6. 卫生间

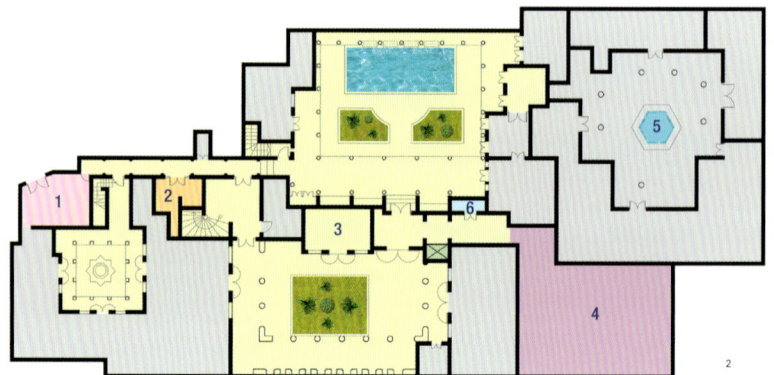

4. Spa massage area
    水疗中心按摩区
5. Boutique
    精品店
6、7. Details
    细部
8. Hamman
    哈曼浴室
9. Bathroom
    浴室
10. Spa products
    水疗产品

11. Restaurant
    餐厅
12. Deluxe suite
    奢华套房
13. Prestige room
    顶级客房
14. Details
    细部

Grand Hotel Bahia del Duque Resort

巴伊亚公爵度假酒店

Completion date: 2008
Designer: Andres Pineiro, Pascua Ortega
Location: Costa Adeje, Tenerife, Spain
Photographer: Grand Hotel Bahia del Duque Resort
Area: 100,000sqm

竣工时间：2008年
项目地点：西班牙，特内里费岛
设计师：安德列斯·皮内罗；帕斯库亚·奥特加
摄影师：巴伊亚公爵度假酒店
项目面积：100,000平方米

The Gran Hotel Bahía del Duque is situated on an extensive estate covering 100,000 square metres in a bay called 'Bahia del Duque', on the shoreline of the municipal district of Adeje in the south of the island of Tenerife. This work designed by the prestigious architect Andrés Pineiro, recreates a small village, with nineteen low density independent houses, breaking away from the concept of a tall, voluminous building. With a total of 357 rooms, 46 of which are suites, the hotel is a faithful reflection of the early 20th century Canary Island architecture. Within this luxury resort, there are four highly exclusive areas.

The Villas del Duque, opened in January 2008, were designed by prestigious interior designer Pascua Ortega, who was helped with the collaboration and supervision by the Canary Island decorator Perico Mesa. These 40 villas are now a sound benchmark of the most exquisite luxury. Outside, a private infinity style pool, gardens of tropical and sub-tropical flora; and inside, bedrooms and bathrooms finished in basalt stone and vertical overhead lighting, aromatic candles and amenities from the luxurious Italian brand Acqua di Parma and sound equipment by Bang & Olufsen. These exclusive residences offer extraordinary services including dinners prepared by a private chef served in the villa, an extensive menu of aroma-therapy baths (romantic, energetic, relaxing, etc.) that are prepared by the team that works exclusively for the Villas, massages inside or outside provided by a team of therapists from the Spa, personalised breakfasts, etc. Moreover, this exclusive area of Villas has its own reading and tea lounge with several kinds of tea that are served all afternoon, snack bar and a spectacular common pool with a la carte restaurant service.

巴伊亚公爵度假酒店占地面积100,000平方米，坐落在名为"巴伊亚公爵"的海湾上，位于特内里费岛的南部。著名建筑师安德列斯·皮内罗打破了高层建筑的设计理念，将其打造成为一个由19座低密度独立住宅组成的小村庄。酒店共有357套客房，其中46套为套房，真实地反映了20世纪早期加那利群岛的建筑风格。这个奢华的度假村分为四个专属区域。

公爵别墅开放于2008年1月，由知名室内设计师帕斯库亚·奥特加和加那利群岛装潢设计师佩里科·麦萨联合设计。这40座别墅是极致奢华的有力标杆。别墅外，是私人无边界式泳池、热带亚热带花园；别墅内，卧室和浴室都装饰着玄武岩、高架照明灯、芳香蜡烛、意大利奢侈品牌彭玛之源的产品和邦·奥陆芬音响设备。这些专属住宅将为宾客提供非凡的服务，如私人厨师亲自准备晚餐、各种各样的芳香疗法浸浴（浪漫、活力、放松等）、室内外按摩（由温泉水疗中心的治疗师团队提供）、个性化的早餐等。此外，别墅区还设有独立的阅读和品茶室（在午后提供多种茶点）、小吃吧、公共泳池和点餐服务。

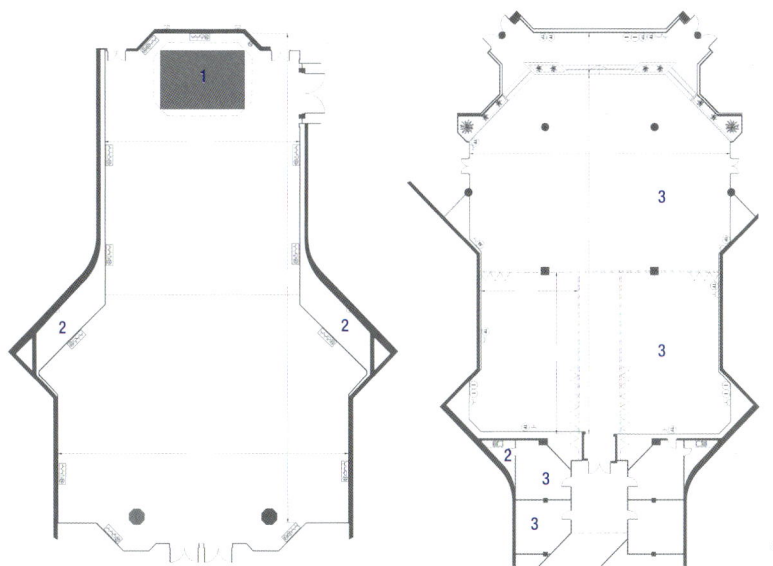

1. Spa outside
   水疗中心外观
2. Spa solarium
   水疗中心日光浴
3. Lobby lounge floor plan
   大堂吧楼层平面图

1. Platform   1. 跳水跳台
2. Office   2. 办公室
3. Salon   3. 沙龙

4. Thalassotherapy circuit pool
   海水浴疗环形池

5. Spa outside cabins
   水疗中心户外小屋

6. Spa relaxation area
   水疗中心休息区

7. Finnish sauna
   芬兰桑拿

8. Spa chromatherapy
   水疗中心色光疗法

9. Kinesis room
   运动室

10. Spa treatment room
    水疗中心治疗室

11. Spa bathroom
    水疗中心盥洗室

12. La Trattoria Restaurant
    La Trattoria餐厅
13. La Tasca Restaurant
    La Tasca餐厅
14、15. Las Aguas Restaurant
    Las Aguas餐厅

Palazzo Arzaga Hotel
Spa & Golf Resort

阿尔扎加宫高尔夫水疗度假酒店

Completion date: 2008 (renovation)
Location: Lake Garda, Italy
Designer: Arch. Silva Giannini (spa area design)
Jack Nicklaus II and Gary Player (Golf Course Design)
Photographer: Pier Paolo Metelli and Chapman Brown

竣工时间：2008年（翻新）
项目地点：意大利，加尔达湖
设计师：艾克·希尔瓦·詹尼尼（水疗区设计）
　　　　杰克·尼克劳斯二世与盖理·普莱尔（高尔夫球场设计）
摄影师：皮尔·保罗·梅泰利与查普曼·布朗

Palazzo Arzaga is a 15th-century mansion harmoniously converted into one of the finest Golf Resorts in Europe. This luxury SPA & Golf Hotel is an exceptional destination for relaxation and pure pleasure, offering all-round comfort and style, just few minutes from Lake Garda.

The hotel adopts the classical interior design, which uses wooden beam ceilings and maintains the original frescos and antiques. It features 84 hotel rooms and suites, and meeting rooms to accommodate up to 180 people. Most rooms have a view over the golf courses or the surrounding countryside. The collection

of Fresco Rooms and Suites preserves the original frescoes of the 15th century and enjoys stunning views of the Arzaga Golf Courses. The new soft high quality bedding gives a fresh look to the rooms. Il Moretto is the Hotel's gourmet restaurant serving fine Italian cuisine accompanied by fine wines from local vineyards. Palazzo Arzaga Golf Courses have been meticulously designed by Jack Nicklaus II and Gary Player to satisfy both beginners and keen golfers.

Arzaga SPA restyles the attractive new design and personalised services. It is an exclusive place of pure relaxation and pleasure for the 5 senses; it is a garden of tranquility and elegance, to live an unique and sensorial experience. The Arzaga's well-being experts will drive the guests along the most suitable path to achieve their balance: the Path of Clouds, the Path of Country, the Path of Flame (SPA Suite), the Path of Water. More facilities include outdoor pool, fitness centre, tennis courts and jogging track, immersed in the 1, 477, 103 square metres of natural reserve surrounding the Resort.

The features and novelties of this timeless place are its 'Paths':

- The Path of Clouds: relaxation path consists of a

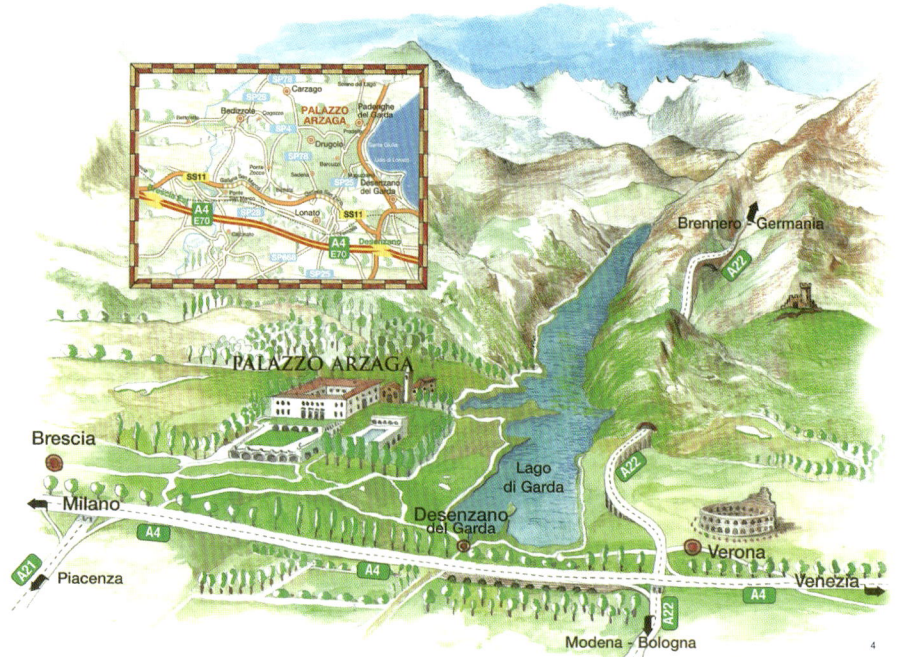

GPS: 45.5112°N 10.4784°E

low temperature sauna with essential oils, a Finnish sauna, a steam bath, showers and the cloud room where you can relax yourself and enjoy a warm and regenerating herbal tea.

- The Path of Country: where you can choose from a wide range of manual treatments.
- The Path of Flame: the Spa Suite dedicated to personalised and exclusive paths to celebrate important recurrences or simply to share a moment with your partner or friends. The route includes sauna, steam bath, emotional shower, whirlpool bath and relaxation accompanied by tea customisable through a variety of herbs and flowers.

阿尔扎加宫度假酒店，是由15世纪的宅邸改造而成的欧洲最好的高尔夫度假村之一。这座奢华的高尔夫水疗度假酒店距加尔达湖仅几分钟车程，提供全方面、各种风格的舒适服务，是休闲娱乐的绝佳之地。

采用古典的室内设计，设有木质横梁天花板、原来的壁画和古董家具。酒店提供84间特色标准客房与套房，还有可以容纳180人的大会议室。大部分客房都可以欣赏高尔夫球场或周边郊外的景色。壁画客房与套房保存了原来15世纪的壁画，还可以欣赏阿尔扎加高尔夫球场的绝美景色。重新更换的高质量床品让客房焕然一新。酒店餐厅Il Moretto为客人提供意式美食还有当地特产的葡萄酒。由杰克·尼克劳斯二世与盖理·普莱尔精心设计的阿尔扎加宫高尔夫球场，既可以满足初学者，也可以满足高尔夫球者的要求。

阿尔扎加水疗馆重塑了新颖的迷人设计，并提供个性化的服务。这里是人们全身心放松与娱乐的专属之地，是一个宁静而典雅的花园，在这里可以享受独特的感官体验。阿尔扎加的健康专家会为客人选择可以达到他们身体平衡的最佳方案：云径—乡村径—火焰径（水疗套房）—水径。其他设施还有户外游泳池、健身中心、网球场和慢跑跑道，淹没于度假村周围1477103平方米的自然保护区内。

这个永恒之地的特色与新奇之处是它的各个"路径"。

云径：是通过一种低温桑拿进行放松的途径，这里有精油、芬兰桑拿、蒸汽浴、淋浴和云室，在这里您可以充分放松自己，还可以享用一杯提神的热花草茶。

乡村径：这里有各种不同的人工治疗方法供您选择。

火焰径：这里的水疗套房是庆祝重生或者与家人朋友享受美好时光的个性而专属之地。这里有桑拿、蒸汽浴、情感淋浴、漩涡浴，还有可以享用各种花草茶的休息区。

9. Spa Suite detail
   水疗套房细部
10. Spa relaxation area
   水疗中心休息区
11. Spa Suite
   水疗套房
12. Fresco room
   壁画客房
13. Classic Residenza room
   经典住宅客房
14. Wine cellar
   酒窖
15. Suite 122 bedroom
   122套房卧室
16. Suite 122 living room
   122套房客厅
17. Classic Palazzo room
   经典宫殿客房

Verdura Golf & Spa Resort

佛杜拉高尔夫水疗度假村

Completion date: 2009 (renovation)
Location: Sicily, Italy
Designer: asa studioalbanese (Achitecture)
Olga Polizzi and Flavio Albanese (Interior Design)
Photographers: Germano Borrelli, Roberto Patti,
Adrian Houston
Area: 2,300,000sqm

竣工时间：2009年（翻新）
项目地点：意大利，西西里岛
设计师：asa studioalbanese建筑事务所（建筑设计）
奥尔加·伯利兹和弗拉维奥·阿尔巴尼斯（室内设计）
摄影师：尔马诺·波莱利，罗伯托·帕蒂，爱德里安·休斯顿
项目面积：2,300,000平方米

The 'Verdura Golf & Spa Resort' is a scheme to a luxury tourist complex for Rocco Forte Hotels on the southern coast of Sicily close to Sciacca, 30 km from Agrigento. The area concerned consists of 2,300,000 square metres of Mediterranean countryside facing the crystalline waters of Sicily Sea.

The scheme is characterised by a respectful and minimal exploitation of the land, emphasising building quality and a virtuous relationship between buildings and the environment. The plan includes 56 luxury villas conceived as a spontaneous aggregation of houses plunged into the Mediterranean landscape

that recalls the traditional farm buildings of the region, a hotel of 200 rooms varying in size and level, conceived in two architecturally different parts: a SPA made up of 12 pavilions and a business centre.

In terms of form, the resort tends to a contemporary architectural statement in keeping with the Mediterranean atmosphere, supporting a new concept of genius loci. Clean lines and pure volumes coexist with materials (volcanic dry-stone walling and rendered surfaces) and artisanal local tradition techniques. All the structures are conceived to be permeable by the sky and in relation with the sea, integrated with the natural environment and integrating it into its own volumes.

The resort is structured by an epicentral building containing all the common spaces (lobby, reception, lounge, bars, restaurants) and different room structures of various types, on the beach or along the downward slope of the landscape. Every room, in the spirit of local tradition, features internal patios or private terraces. The President Suite category instead contains its own pool and hydromassage.

The Verdura Spa is set apart from the activity of the resort, featuring 4,000sqm pristine space. The

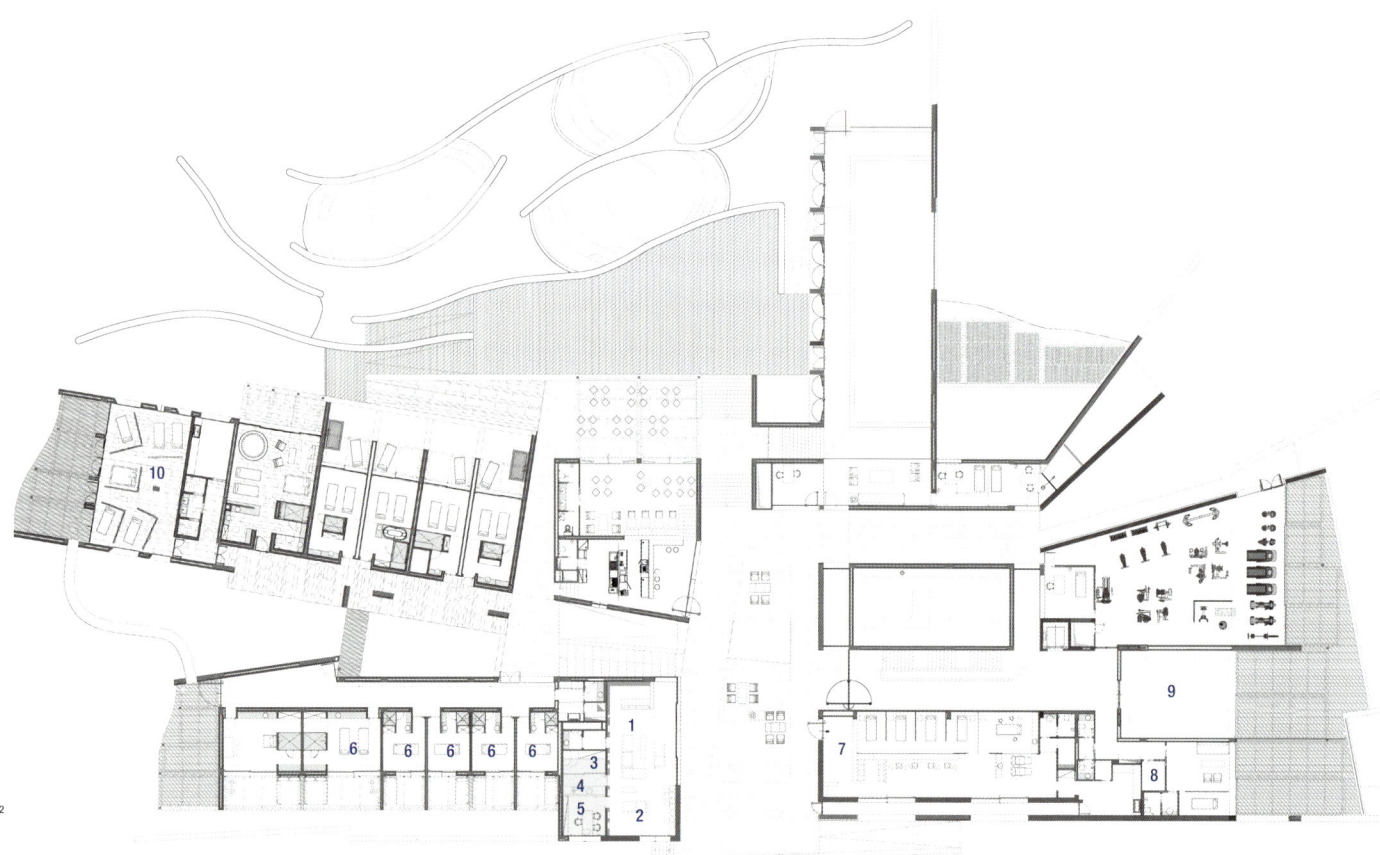

1. Reception
2. Shop
3. Store
4. Pantry
5. Office
6. Treatment room
7. Beauty salon
8. Yoga store
9. Yoga room
10. Relaxation area

1. 接待处
2. 商店
3. 储藏室
4. 餐具室
5. 办公室
6. 治疗室
7. 美容沙龙
8. 瑜伽用品商店
9. 瑜伽室
10. 休息区

conference and event space is private, flexible and designed to perform as a creativity centre or a place to have party. The project also incorporates solar energy and water-recycling, a focal point in Sicily where rainfall is minimal and concentrated around short periods in the year.

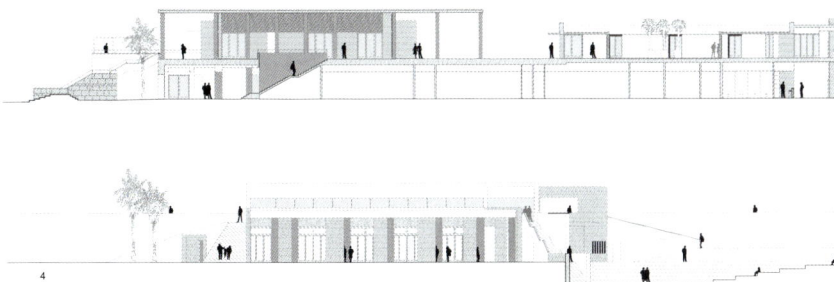

佛杜拉高尔夫水疗度假村是洛克福特酒店集团一项奢华旅游度假区计划，位于西西里岛南海岸，临近夏卡，距离阿格里真托市30公里。该度假村包含2300000平方米的地中海乡村，面朝如水晶般透明的西西里海。

该计划的特点是充分尊重并且最小限度地开发这块土地，重视建筑质量和建筑与环境的良性关系。该计划包括56套奢华别墅，设计构思是建立一个融入地中海景观的自发性住宅群，可以让人联想到该地区传统的农业建筑。还包括一的酒店建筑，拥有200间不同规模与级别的客房，该建筑设计构思是将其分为两个建筑结构不同的部分：由12个凉亭构成的水疗中心和会议中心。

在建筑形式上，该度假村趋向于符合地中海氛围的当代建筑风格，是当地一个全新的理念。简洁的流线和纯体量与各种材料（火山岩构造的墙体和渲染的墙面）和当地传统技术的手工艺品并存。所有结构的构思是要达到与天相接，与海相连，与自然环境相融合，并将所有这些元素都整合到自己的体量中。

该度假村由一栋中心建筑和各种类型的客房建筑构成，其中中心建筑包含所有公共空间（大堂、接待处、休闲吧、酒吧、餐厅），各种客房建筑位于海滨及景观下坡处。每间具有当地传统特色的客房都有室内露台和私人阳台。总统套房设有私人泳池和水疗按摩设施。

佛杜拉水疗中心与度假村的其他活动分开，占有4000平方米的原始空间。会议室与活动室的设计具有私密性，也是灵活多用的，可以作为创意中心或者举办聚会的场所。项目设计上还利用了太阳能和水循环系统，这在西西里地区是非常必要的，因为这里降雨量很小，并且集中在一年中很短的一段时间里。

1. Spa & Thalassotherapy pools
   水疗中心与海水浴理疗池
2. Resort plan
   度假村平面图
3. Spa entrance lounge
   水疗中心入口休息区
4. Centre facilities elevation
   中心设施立面图
5. Resort pool
   度假村泳池
6. Presidential suite pool
   总统套房泳池
7. Resort reception
   度假村接待处
8. Resort reception lobby
   度假村接待处大堂
9. Spa section
   水疗中心剖面图

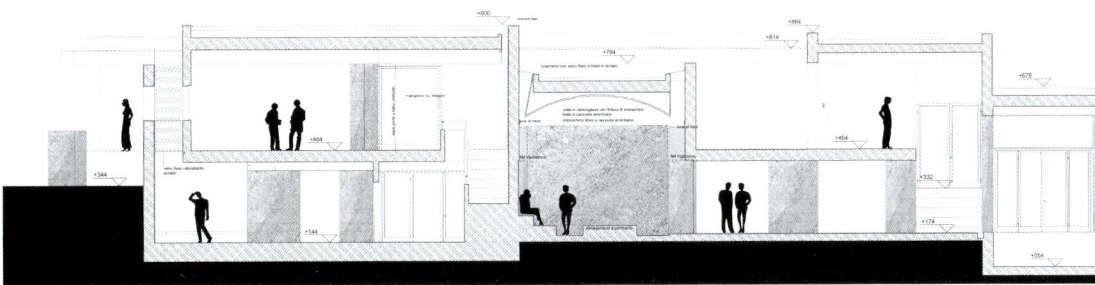

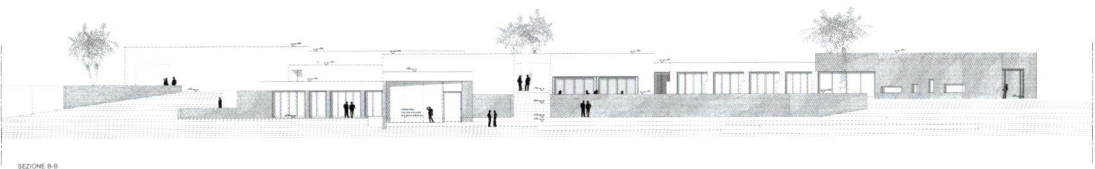

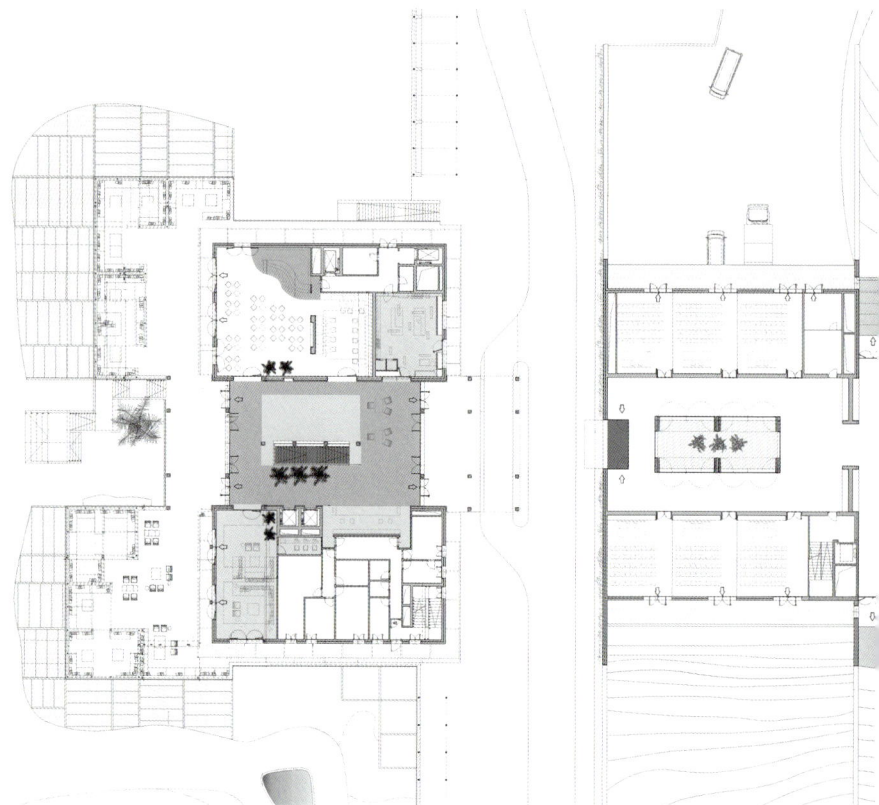

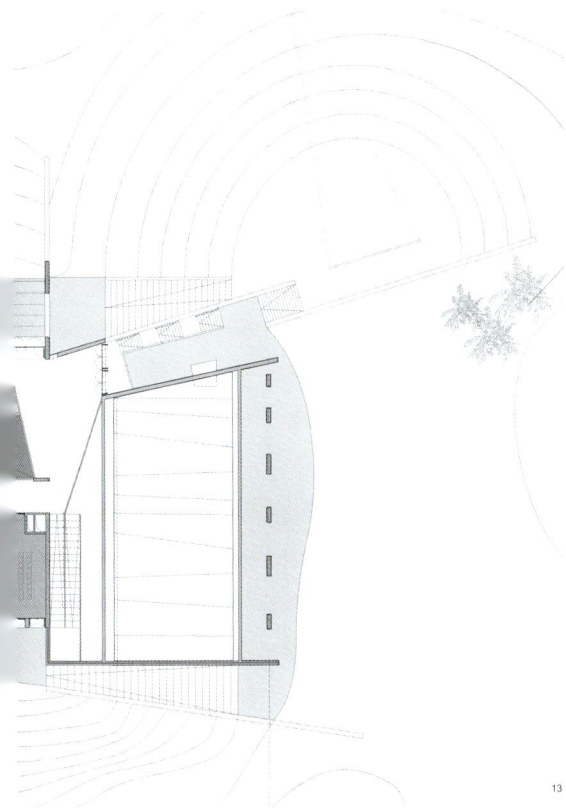

Grand Resort Lagonissi

拉格尼西豪华度假村

Completion date: 2010 (last renovation)
Location: Athens, Greece
Designer: ARCH Group Architects, Davide
Macullo Architects
Photographer: George Augustinatos
Area: 291, 000sqm

竣工时间：2010年（翻新）
项目地点：希腊，雅典
设计师：ARCH Group 建筑师事务所，Davide
Macullo建筑师事务所
摄影师：乔治·奥古斯汀那特斯
项目面积：291, 000平方米

Grand Resort Lagonissi is a secluded island resort in the heart of the Athenian Riviera, where the sea is inviting and sparkles in the glowing sunlight. Its impeccable facilities are found in an outstanding choice of accommodation which offers the utmost in luxury design and high-tech amenities from spacious waterfront rooms to top-end Platinum Club villas with their own terraces, gardens, pools, personal chef and butler.

The elegant Penthouse Suites, which are located at the top floor, offering privileged view of the Mediterraneo bay through their floor-to-ceiling windows, are also

equipped with the latest technological devices. The spacious sea view bathroom is equipped with an aero spa bath-tub, a separate steam bath cabin decorated with fine petit mosaic.

The accommodation in the Belvedere Suite offers an extensive variety of choices featuring a high-tech approach, such as remote controlled mattresses, air bathtub and Jacuzzi, steam bath cabin. The Deluxe Suite on the waterfront is designed to satisfy the highest expectations. The bungalows and suites provide a separate work-out area with top-of-the-line exercise equipment.

The Governor Villa is a seafront heavenly escape which has refined ambience. The open plan style opulent bedrooms of the Residence Villa are adorned with sumptuous oversized beds and lavish interiors all made of oak. Elegant fabrics in soothing colours complement the furnishings combined with contemporary art deco touches. The stunning sea view is accessible even from the deluxe Jacuzzi bathtub.

Hopelessly romantic, the unique Dream Suite is a luxurious hideaway for those who desire a delightful extravagance. Luxury and technology blended harmoniously, providing outstanding amenities

1. Veghera club
2. Ouzeri Aegeon Greek Restaurant
3. Captain's House Italian Restaurant
Captain's Deck bar
4. Helipad

5. Thalaspa Chenot
6. Water sports
7. Mediterraneo Restaurant
8. Veranda bar
9. Aphrodite restaurant
10. La Piscina Poolside bar
11. Indoor parking

12. The Grand Hall
13. Kohylia Polynesian restaurant
14. Kids Club
15. Grand Pier
16. Grand Beach, Water sports

17. Poseidon Restaurant

1. Veghera俱乐部
2. Ouzeri Aegeon希腊餐厅
3. 船长屋意大利餐厅
船长甲板酒吧
4. 直升机场

5. Thalaspa Chenot水疗中心
6. 水上运动中心
7. 地中海餐厅
8. 游廊酒吧
9. 阿佛洛蒂特餐厅
10. 泳池边酒吧
11. 室内停车场

12. 大厅
13. 科里亚波利尼西亚餐厅
14. 儿童俱乐部
15. 码头
16. 大型海滩水上运动中心
17. 波塞冬餐厅

such as control panel for devices, designer furniture, large outdoor heated pool and private gym. The remote controlled retractable skylight, above the king size bed, reveals bright blue morning skies and romantic night stars. The crown jewel of the resort, the Royal Villa offers the ultimate in lavish exclusivity, occupying its own private cove by the water edge. The indoor pool quarters consists of a gym with work-out area, a large heated pool with reverse mode system and hydro massage device as well as a separate steam bath and a massage room. The fully equipped business centre includes all modern facilities.

The Grand Resort Lagonissi extends natural beauty in its design and offers the finest contemporary luxuries and services. The resort offers six waterfront restaurants perfect for lunch and dinner, and four bars admiring the view. Through the exclusive Thalaspa Chenot treatments and the unique seafront fitness centre, guests achieve an internal balance and harmony. Surrender your senses to the lavish world of Thalaspa Chenot. Oriental therapies with natural spices and scented flowers are incorporated to invigorate your body and mind.

拉格尼西豪华度假村位于雅典河湾中心，是一个隐蔽的岛上度假村，这里的海水在炎热的阳光下闪闪发光，充满诱惑。客房服务设施齐全，客房类型多样，设计奢华，配有高科技的服务设施。其中有宽敞的海滨客房，还有顶级白金俱乐部别墅，别墅有各自的阳台、花园、游泳池、私人厨师和管家。

典雅的顶级公寓套房位于酒店的顶层，配有最新的技术设备，可以通过落地窗欣赏特有的约翰湾美景。宽敞的海景卧室配有一个水疗浴缸和一个独立的以马赛克装饰的蒸汽房。

观景套房提供各种各样的高科技设备，例如遥控床垫、透气浴缸与极可意水流按摩浴缸和蒸汽房。海滨豪华套房的设计可以满足最高需求的客人。平房和套房提供独立的带有顶级运动设施的健身房。

州长别墅具有一种优雅的氛围，是一个海滨避世天堂。住宅别墅内的客房是开放式的布置风格，装有华丽的大床和各种由橡木制成的室内装饰。典雅的纤维织物，色彩柔和，与结合了当代装饰艺术风格的家具和谐一致。即使在奢华的极可意水流按摩浴缸里也可以欣赏绝妙的海景。

独特的梦想套房，极具浪漫氛围，是追求奢侈与娱乐的豪华胜地。奢华与技术相融合，提供各种高级的服务设施，例如各种设备的控制面板，专门设计师设计的家具，大的户外热水游泳池和私人健身房。超大号床的上方有可遥控的伸缩天窗，可以在床上欣赏早上蔚蓝的天空和晚上浪漫的星空。

皇家别墅是度假村里的皇冠之珠，这里独占一个私人水湾，极具奢华与专属性。室内泳池区包括一个健身房和一个大的热水游泳池，游泳池装有逆向模式系统和水疗按摩设施，还有独立的蒸汽浴室和按摩室。设备齐全的会议中心包含所有现代服务设施。

拉格尼西豪华度假村在设计上使天然的美景得到延伸，并提供了当代的奢华与服务。度假村有6间享用午餐和晚餐的海滨餐厅和4间酒吧，景色迷人。在专属的Thalaspa Chenot水疗中心和独特的海滨健身中心，客人们可以使身体内达到均衡与调和。在Thalaspa Chenot这个奢华的水疗世界，可以使你得到全身心的放松，使用东方治疗方法，结合天然香料和芳香的鲜花重新激活你的身体和心灵。

1. Mediterraneo Restaurant
   地中海餐厅
2. Resort plan
   度假村平面图
3. Captain's House outdoor decks
   船长餐厅室外甲板区
4. Wedding setup
   婚礼设施

5. Thalaspa Chenot reception area
   Thalaspa Chenot水疗中心接待区
6. Thalaspa Chenot manicure-pedicure area
   Thalaspa Chenot水疗中心修甲区
7. Thalaspa Chenot hydrotherapy room
   Thalaspa Chenot水疗中心水疗室
8. Royal Villa outdoor heated pool
   皇家别墅户外加热泳池
9. The Dream Villa terrace
   梦想别墅露台
10. Governor two-bedroom Villa with private pool
    带私人泳池的两居室州长别墅

Mezzatorre Resort & Spa

麦择特瑞水疗度假村

Completion date: 2008 (renovation)
Location: Ischia, Italy
Designer: Studio Izzo & Partners
Photographer: Mezzatorre Resort & Spa
Area: 70,000sqm

竣工时间：2008年（翻新）
项目地点：意大利，伊斯基亚
设计师：Izzo & Partners设计工作室
摄影师：麦择特瑞水疗度假村
项目面积：70,000平方米

In the Isle of Ischia, surrounded by 70, 000 square metres of pine wood, in one of the most beautiful and still unsullied corners of Mediterranean maquis, very close to the magnificent villa that belonged to Luchino Visconti, a famous Italian film director, there is the Mezzatorre Resort & Spa. It sits at the peak of a promontory rising steeply out of the sea, a unique position that makes it a haven of peace and quiet.

The main body is an old tower dating back to the 16th century that belonged to Baron Fassini and was used by the local residents to defend themselves against the Saracens; the other buildings are scattered among the

pines and holm oaks encircling the hotel, and blend perfectly with the landscape and the style of the tower. The hotel's nice terraces are the ideal place to enjoy breath-taking sunset, indulge in your favourite readings and have an unforgettable time accompanied by romantic background music or relish the typical Mediterranean dishes served at the two restaurants.

If you are looking for a cosy and elegant place to spend your holiday and restore your body and spirit, this exclusive hotel with rooms and suites facing the sea and the greenery is for you. If you want to benefit from the well-known thermal water and muds of the island, the well-trained staff of the very modern Spa is going to advise you the most updated technologies of balneotherapy. The Spa includes also a Health & Beauty Centre where a wide range of beauty treatments is carried out under the supervision of highly qualified physicians.

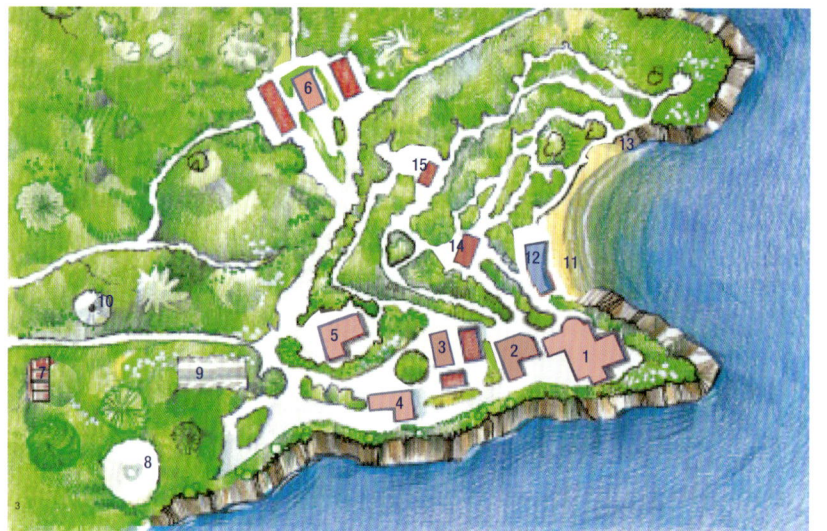

1. Torre Rooms and Spa	1. 托雷客房与水疗室
2. Restaurant	2. 餐厅
3. Serra Rooms	3. 萨拉客房
4. Reception & Rooms	4. 接待处与客房
5. Foresteria Rooms	5. 弗雷斯塔利亚客房
6. Visconti Rooms	6. 维斯康蒂客房
7. Tennis	7. 网球场
8. Heli-pad	8. 直升机场
9. Parking	9. 停车场
10. Chapel	10. 教堂
11. Landing-stage	11. 码头
12. Sea water swimming pool	12. 海水游泳池
13. Private pathway to the sea	13. 通往大海的私人小径
14. Gazebo Lounge Bar	14. 露台休闲吧
15. Deluxe Bungalow Suite	15. 奢华别墅套房

麦择特瑞水疗度假村位于伊斯基亚岛，地中海马基群落最美的一块净土上，被70000平方米的松木所环绕，该度假村离意大利著名电影导演卢奇诺·维斯康蒂宏伟的别墅非常近。度假村建在陡峭的海角顶部，独特的地理位置使它成为一个和平与宁静的避风港。主建筑是可以追溯到16世纪的古塔建筑，原属于Fassini男爵，曾被当地人用于抵御萨尔逊人；其他的建筑掩映于松树和圣栎间，围绕在酒店主建筑周围，与周围的景观和主塔建筑的风格完美的融合。

酒店漂亮的露台是欣赏落日时分绝美景色的理想之地，还可以沉浸于你所喜爱的读物，在浪漫的音乐中品尝地中海特色小吃，度过令人难忘的时光。

如果你在寻找一个舒适而高档的地方度假或放松心灵，这个拥有海滨客房与套房的专属酒店是你最佳的选择。如果你想在岛上著名的热水和泥浆中放松身心，现代风格的水疗中心里训练有素的理疗师将为你提供最先进技术的治疗。水疗中心里还有一个健身与美容中心，这里高水准的美容顾问会为你提供一系列的美容方案供您选择。

1. Outdoor heated swimming pool and poolside restaurant
   户外加热游泳池与池边餐厅
2. Resort aerial view
   度假村鸟瞰图
3. Resort plan
   度假村平面图
4. Deluxe Grand Suite terrace with Jacuzzi
   带极可意水流按摩浴缸的豪华套房露台
5. Outdoor massage area
   户外按摩区
6. Interior pool
   室内泳池
7. Bungalow Family Suite terrace
   别墅家庭套房露台
8、9. Gazebo Piano Bar
   露台钢琴酒吧
10. Bungalow Family Suite sitting room
    别墅家庭套房起居室
11. Lobby corner
    大堂一角
12. Bungalow Family Suite bedroom with canopy bed
    别墅家庭套房内带天篷床的卧室

Capella Ixtapa

伊斯塔帕嘉佩乐度假村

Completion date: 2008
Location: Ixtapa, Mexico
Designer: Enrique Muller and Santiago Aspe
Photographer: Robert Reck

竣工时间：2008年
项目地点：墨西哥，伊斯塔帕
设计师：恩里克·穆勒和圣地亚哥·阿斯匹
摄影师：罗伯特·莱克

Between Mexico's Sierra Madre Mountains and the Pacific Ocean, surrounded by lush, tropical forests lies Capella Ixtapa, a secluded, romantic and exclusive resort whose location is so private it's like sharing a well-kept secret. Set alongside a stunning rock cliff with endless views of the breathtaking azure blue ocean, Capella Ixtapa offers its guests a hideaway setting that belies the wealth of attractions just minutes away.

The resort's design celebrates Mexican culture with a sophisticated blend of modern and traditional design elements.

Ixtapa-Zihuatanejo is nestled on the Pacific Coast, 152 miles northwest of Acapulco and is part of the 'triangulo del sol,' or triangle of the sun, of the Guerrero state in the southern meridional region of Mexico. The Nahauatl dialect 'Ixtapa' means 'the white place' in reference to the pristine white sands of the area. Once a coconut plantation close to the quaint fishing village of Zihuatanejo, Ixtapa is now an intimately modern, environmental-friendly resort area ideal for travellers who prefer small, secluded destinations to bustling cities and major destinations. Ixtapa-Zihuatanejo continues to impress with its natural attributes – from sandy beaches and rugged mountains to exuberant vegetation – setting the stage for some of the finest sport fishing, golf and eco-sports available anywhere. Additionally, the protected waters surrounding Capella Ixtapa are home to a variety of sea life and vegetation.

Capella Ixtapa is located in proximity to the world-class Marina Ixtapa and two championship golf courses. For guests seeking an authentic Mexican experience, the local markets at Zihuatanejo are only ten minutes away. A variety of local shops, dining and entertainment are also within easy reach.

1. Bedroom	14. Water	1. 客房	14. 水池区
2. Swimming pool	15. Pool terrace	2. 游泳池	15. 泳池露台
3. Solarium	16. Beach	3. 日光浴室	16. 海滩
4. Palapa	17. Sundeck	4. 简陋草屋	17. 日光浴处所
5. Dining pergola	18. Terrace	5. 就餐绿廊	18. 阳台
6. Exit	19. Reception Bar	6. 出口	19. 接待酒吧
7. Entry	20. Restaurant	7. 入口	20. 餐厅
8. Bridge	21. Men's toilet	8. 架桥	21. 男卫生间
9. Gazebo	22. Women's toilet	9. 露台	22. 女卫生间
10. Laundry	23. Elevator	10. 洗衣房	23. 电梯
11. Lobby	24. Multiple uses	11. 大厅	24. 多功能区
12. Access to bathrooms	25. Employee dining	12. 浴室入口	25. 员工餐厅
13. Deck	26. Slope	13. 甲板	26. 斜坡

Capella Ixtapa extends over Don Juan beach, a small stretch of the Pacific coastline, and offers guests the utmost privacy in a stunning environment. Each of the 59 guest rooms cascade down a cliff towards the sea, boasting breathtaking views of the Pacific Ocean. Each guestroom features a large, private outdoor terrace and an individual plunge pool thoughtfully positioned out of sight from other suites to ensure privacy.

1. Resort exterior at night
   度假村夜景

2. Resort plan
   度假村平面图

3. Tappas bar terrace
   Tappas酒吧露台

4. Sea view dining
   海景餐厅

5. Las Rocas Restaurant
   Las Rocas餐厅

6. Wedding set up
   婚礼设施

伊斯塔帕嘉佩乐度假村坐落在墨西哥马德雷山脉和浩瀚的太平洋之间，四周环绕着茂密的热带森林。这座诱人而浪漫的度假村的位置是如此私密，仿佛与人共享着不为人知的秘密。伊斯塔帕嘉佩乐度假村沿着绝妙的悬崖峭壁而建，享有碧蓝色的海洋的不尽美景，为宾客提供了一个世外桃源。

度假村的设计崇尚墨西哥文化，同时又巧妙地融合了现代和传统元素。

伊斯塔帕－芝华塔内欧地处太平洋海岸，阿卡普尔科西北152英里处，是墨西哥南欧区域格雷罗省"太阳三角"的一部分。"伊斯塔帕"在当地语言中意为"白色之所"，暗喻该地区纯白的沙子。伊斯塔帕曾是芝华塔内欧渔村边上的一个椰子园，而现在却被打造成了私密、现代而环保的度假村，适合钟爱小型度假之地的游客从喧嚣的都市和大型景点中前来寻求平静。沙滩、起伏的山脉、茂密的植物……这些元素都让前往伊斯塔帕－芝华塔内欧的游客惊喜万分，也为进行钓鱼活动、高尔夫和其他体育运动提供了理想的舞台。此外，伊斯塔帕嘉佩乐度假村四周受保护的水域还是各类海洋动植物的天堂。

伊斯塔帕嘉佩乐度假村紧邻世界级码头伊斯塔帕和两座锦标赛高尔夫球场。对于想体会墨西哥文化真谛的游客，芝华塔内欧的本地市场距离度假村仅有10分钟的路程。各种本地商店、餐饮娱乐设施也都相距不远。

伊斯塔帕嘉佩乐度假村一直延伸到唐璜海滩——太平洋海岸线的一个分支，为宾客们在绝妙的景色中提供极致私密之所。59套客房沿着峭壁顺势而下，享有太平洋的绝美景色。每套客房都拥有一个巨大的私人露台和私人游泳池，二者被贴心地设在其他宾客看不到的地方，保证了隐私。

7. Entrance corridor
   入口通道
8. Resort elevation
   度假村立面图
9. Entrance corridor
   入口通道
10. Sea view terrace
    海景露台
11. Poolside cabana
    池边小屋

12. Spa reception
   水疗中心接待处
13. Guest room terrace
   客房露台
14. Spa beauty
   水疗中心美容室
15. Spa treatment room
   水疗中心治疗室
16. Spa pool
   水疗池
17. Spa dressing room
   水疗中心更衣室
18. Building elevation
   建筑立面图

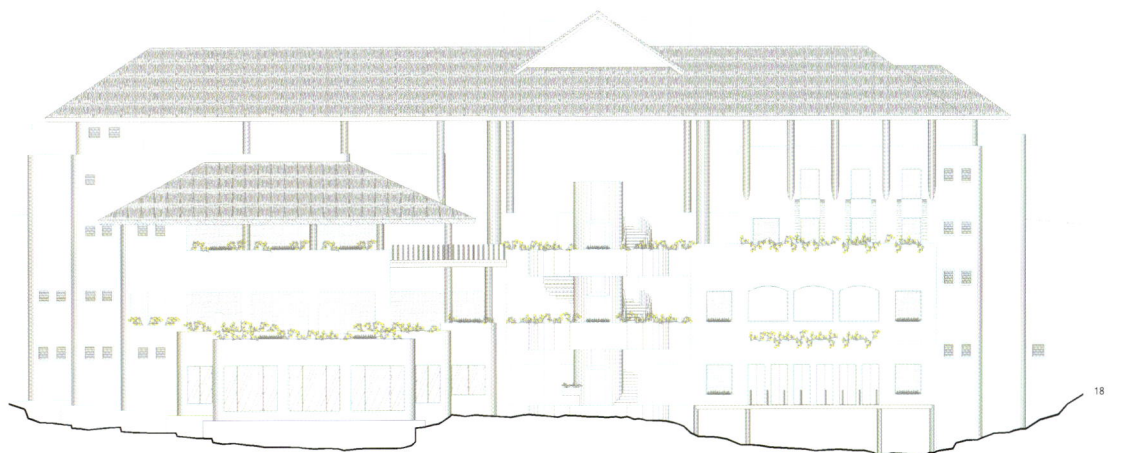

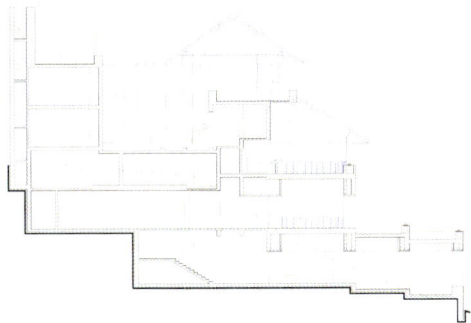

21

22

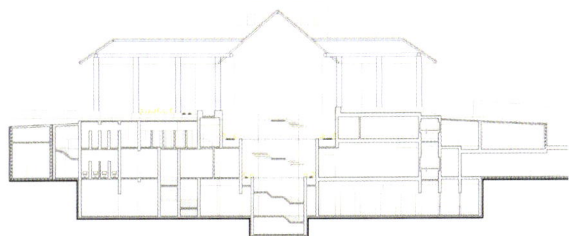

23

24

25

19. Bar
酒吧
20. Amares Restaurant
Amares餐厅
21、22. Building section
建筑剖面图
23、24. Ocean view guest room
海景客房
25. Guest room
客房

Hacienda Tres Ríos Resort, Spa & Nature Park

蔡斯里奥斯庄园水疗度假村与自然公园

Completion date: 2008
Location: Riviera Maya, Mexico
Designer: Sunset World Resorts & Vacation
Experiences leaded by Romarico Arroyo
Photographer: Hacienda Tres Ríos Resort
Area: 1, 315, 228sqm

竣工时间：2008年
项目地点：墨西哥，拉玛雅
设计师：罗马瑞克·阿罗约领导的世界夕阳度假集团与度假体验团队
摄影师：蔡斯里奥斯庄园度假村
项目面积：1, 315, 228平方米

The Hacienda Tres Ríos is an all-inclusive luxury resort on Mexico's Riviera Maya that was designed and built with sustainability in mind. The hotel's designers and architects studied the environment and found ways to preserve and protect local plants and animals while incorporating the best sustainable practices into the construction and operation of the hotel. Hotel guests can enjoy the hotel's beautiful surroundings, attractive rooms and private nature park knowing they are making very little impact on the local environment.

Hacienda Tres Ríos is the only environmentally

responsible luxury resort that combines architecturally acclaimed accommodations, fine dining and tasteful entertainment, a world-class spa and a chance to exercise your choice of sports and recreations at Tres Ríos Nature Park. The resort has 273 suites with breathtaking views of the sea to enjoy. and guests can live the authentic vacation experience while protecting the natural beauty and culture of the Mexican Caribbean.

Consistent with its sustainable mission, Hacienda Tres Ríos has implemented technology and environmental practices from its conception with environmental impact studies which identified the animal and plant species of the area. This was determined that all structures would be prefabricated away from the location and finally assembled in carefully selected areas with low environmental value. To ensure the natural surface water flow into the mangroves, the hotel was erected on a permeable rock base instead of a concrete slab platform.

This resort finds environmental balance, by making responsible and intelligent use of natural resources. The air conditioning system is cooled with cold water pumped from natural wells and flows through

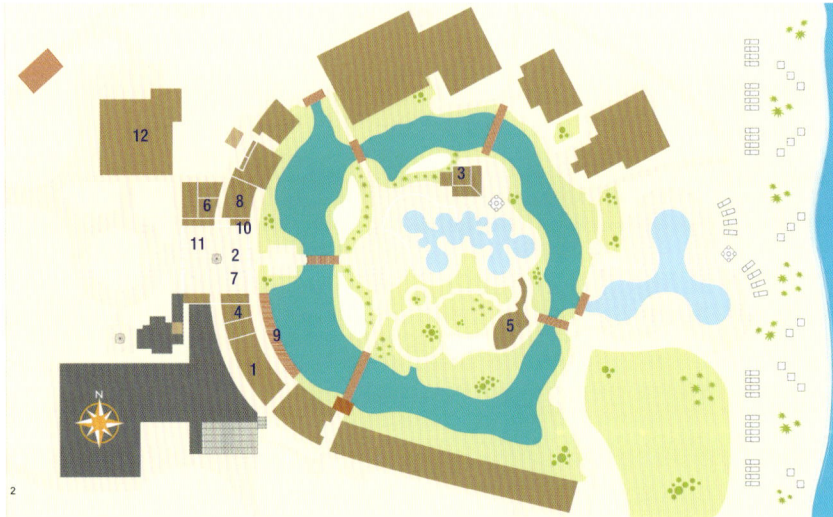

pipes that eventually takes it back to its origin without altering its purity. This procedure reduces energy consumption by 40%. The heat generated is applied to heat water for shower heads, same to which ambient air is injected to increase pressure and save water; a desalination plant was installed to convert salt water into the highest quality, pure, fresh water, thus all the water used in the hotel is desalinated. This also prevents the use of groundwater from the aquifer.

1. Casa Las Islas—International buffet
2. El Alebrije—Fine Mexican
3. Hacienda Grill—Grill & American cuts
4. Kotori & Porto Bello—Alternating Asian and Italian cuisine
5. IL Forno—Pizzeria
6. Patisserie Café de Paris—French pastries, coffee & tea
7. Agave Tequila Bar
8. Victory Sports Bar
9. Martini and Teatime Terrace
10. Lobby Bar
11. Lobby
12. Multifunctional ballroom

1. 拉斯伊拉斯之家——国际风格自助餐厅
2. El Alebrije餐厅——墨西哥风味餐厅
3. 大庄园烧烤——烧烤与美国风味餐厅
4. Kotori & Porto Bello餐厅——亚洲风味与意大利风味餐厅
5. IL Forno比萨店
6. 巴黎蛋糕咖啡厅——法式甜馅饼、咖啡和茶
7. 龙舌兰酒吧
8. 胜利体育酒吧
9. 马提尼和茶酒吧——定时露台
10. 大堂酒吧
11. 大厅
12. 多功能宴会厅

1. Resort aerial view
   度假村鸟瞰图
2. Resort plan
   度假村平面图
3. Building exterior
   建筑外观
4. Poolside restaurant
   池边餐厅

5. Resort lobby
   度假村大堂

6. Kotori Restaurant serving
   Asian gourmet cuisine
   提供亚洲美食的科特瑞餐厅

7. Hacienda Grill
   庄园烧烤餐厅

蔡斯里奥斯庄园度假村是墨西哥拉玛雅地区一家包含所有服务的奢华度假村，设计与建造的过程中都充分考虑了可持续发展性。酒店设计师与建筑师研究了周围的环境并且找到合适的方式来保护当地动植物，同时在酒店建造和经营的过程中也融入了环保理念与措施。酒店的客人在享受酒店周围美景、迷人的客房和私密的自然公园的同时会了解到他们对当地的环境影响不大。

蔡斯里奥斯庄园度假村是唯一一家环保奢华度假村，这里有架构上广受赞誉的客房，精美的餐饮与雅致的娱乐设施，世界级的水疗中心，还有机会在蔡斯里奥斯自然公园里进行各种运动和娱乐活动。该度假村有273套拥有惊人海景的套房，在保护墨西哥加勒比海自然美景和文化的同时，享受真正的度假生活。

以环保为己任，蔡斯里奥斯庄园度假村在设计理念中应用了高科技和环保措施，通过环境影响研究对该地区的动植物种类进行了鉴定。决定将所有的结构远离这个地方建造，最后在精心挑选的地方装配，这样可以降低对环境的破坏。为了确保天然的地表水能够流入红树林中，酒店建在了渗水岩上，而不是混凝土板的平台上。

该度假村合理利用自然资源，寻求环境的平衡。空调系统是通过抽水泵将天然井中的冷水引入到管道中，再从管道流回井中而不会影响水的纯净度，这个程序可以节能降耗40%。产生的热量可以给淋浴用水加热，注入周围的空气可以增加压力进而节约用水。安装了海水淡化装置将盐水转化为高质量的纯净淡水，因此酒店中所有使用的水都是脱盐淡水，这样可以防止使用蓄水层中的地下水。

8. Casa Las Islas Gourmet Restaurant
卡萨拉斯美食餐厅
9. Alebrije Restaurant
埃勒博瑞餐厅
10、11. Casa Grande bedroom
卡萨格兰德卧室
12. Suite couple bedroom
套房双人间卧室

Kenoa – Exclusive Beach Spa & Resort

柯诺亚海滨水疗度假村

Completion date: 2009
Location: Barra de São Miguel, Brazil
Designer: Osvaldo Tenório
Photographer: Rogerio Maranhão
Area: 2,749sqm

竣工时间：2009年
项目地点：巴西，圣米格尔岛
设计师：奥斯瓦尔多·特诺里约
摄影师：罗格瑞奥·马拉尼昂
项目面积：2,749平方米

Like a big reference embroidery in permanent dialogue with nature, the Kenoa Resort project drew itself in Barra de São Miguel, coast of Alagoas, with the signature of architect Osvaldo Tenório. Following ethnic and primitive lines and appropriating a piece of the 256 km of Alagoas blue sea, the space projects itself is like an altar in ode to nature.

The Kenoa is born and extends itself in constant allusion to nature — be it in its recreation or exacerbation of the original landscape and establishing the synthesis between modern, primitive and technology translated into comfort. Backed by the

forest, present in all levels of the structure, the Kenoa Resort is part of an Agora–allusion to the Greek place of assembly/squares. From this symbolic space emerges the whole structure of Kenoa Resort from the SPA, restaurant, leisure area, reception, shop and villas.

The materials reinforce the DNA of Kenoa Resort — its commitment to being natural and environmental friendly — and are applied in new ways and different designs, prompting surprising sensations. The mortar used — shown as a pigmented single-layer — refers to buildings derived from the land. The ethnic and regional players such as stucco and African plaster are softened throughout the project by the amazing wealth of detail in every corner of the resort. Panels of rough golden stone and piassava huts assemble small scenarios, all in ode to the natural and blended with limestone details from Asia and Brasil. Art is remade — as a commitment — to the natural.

In the lobby, a large panel of Indian Tree Bark from Vanuatu Island welcomes the guests. Mining objects from China, Indonesia, Philippines, South Africa, Morocco and the indigenous Brazilian art rebuild the embroidery of references establishing the architect's

intentional congruence of cultures. From there the roots emerge, paths are redone in tributes coming from various parts of the world reflecting the great cultural village inserted in the atmosphere of Alagoas. In the 2,749 square metres of construction, the water serves as a tack, in honor of the state that is signed as Paradise Waters, is the warehouse of them all. On one of the major reflecting pools, blocks of rough stone balance themselves as on an island and in reference to them, become subtle elements.

At Kenoa Resort everything is exact, if unassuming, yet it is exact, the project was born under the sign of surprise and thus is remade in every corner. To translate it well, objects in their natural state, such as rust, arise spontaneously and fall as a natural element from the use and the intention of humanisation of the project, as the great and old pot of iron foundry transformed into a decorative cup, showers in brass, designed by the architect, the taps of oxidized iron, old sugar mills turned into sculpture or the old dormant rail lines that support and fit into the structure, hours show in intention to create new textures and applications.

In the SPA, the architect Osvaldo Tenorio remade his homage to the state and is inspired by the movement of water, turning the place into a temple of relaxation and an encounter to the natural. A large swimming pool and 12 private ones of Kenoa Resort are coated with stones that reproduce the blue colour of the sea and large Jacuzzi also suits the roughness of the stone.

The exclusive and rigorous design, the excellence in detail and commitment with the place is one of the sensations that translate Kenoa Resort. Everything is handmade, unique, valuing the simple and exacerbating form and there resides the translation of Kenoa luxury. All the furniture was especially designed for the project, sofas, tables, chairs, all in contemplation of nature or on behalf of volumetry that offers the ideal contemporaneity to the project.

1. Pool & beach club overview
   泳池与海滩俱乐部全貌
2. Kenoa villa outside view
   柯诺亚别墅外观
3. Kenoa resort entrance
   柯诺亚度假村入口
4. Pool & beach club
   泳池与海滩俱乐部
5. Beach entrance
   通往海滩的入口
6. Resort plan
   度假村平面图

1. Lobby	1. 大厅
2. Reception	2. 接待处
3. Toilets	3. 卫生间
4. Central agora	4. 中央广场
5. Kitchen	5. 厨房
6. Suites	6. 套房
7. Pool deck	7. 泳池甲板区
8. Pool	8. 游泳池
9. Beach deck	9. 海滩甲板区

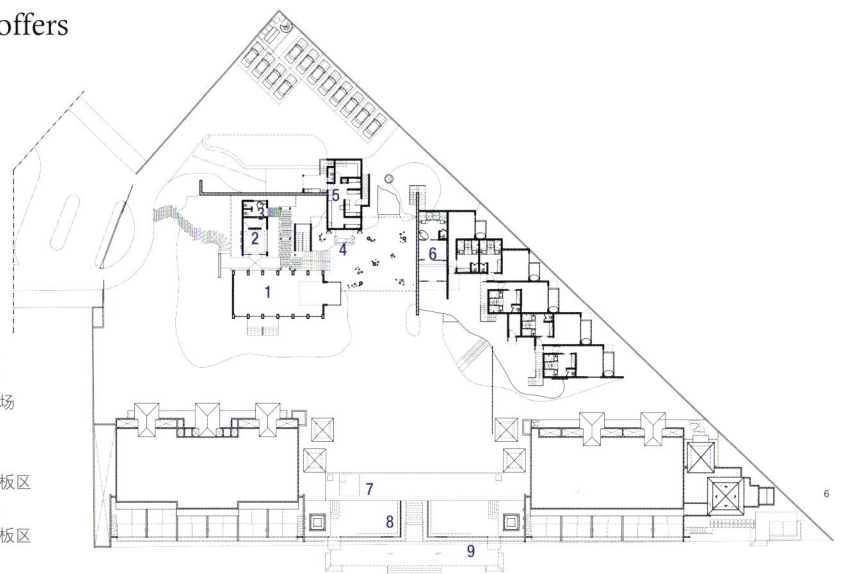

柯诺亚度假村位于阿拉戈斯海岸的圣米格尔岛，由著名建筑师奥斯瓦尔多·特诺里约设计，宛如大自然中一块永恒的织锦。度假村沿着256千米长的蔚蓝海岸线而建，兼具民族和原始风情，无处不在歌颂大自然的美。

柯诺亚度假村与自然紧密相连，再创造并增强了原始景观，形成了融合现代、原始和技术的综合体。在森林的掩映下，度假村仿佛一座广场的一部分，包含温泉馆、餐厅、休闲区、接待区、商店和别墅等多种设施。

材料的运用以自然和环保为主题，以全新的方式和不同的设计凸显了柯诺亚度假村的特色，带来惊喜的感觉。单层彩色灰泥的运用暗指建筑源于土地。灰泥和非洲石膏等民族和当地材料的使用在项目中被大量的细节设计所弱化。粗糙的金石板和棕榈小屋共同组成了小型场景，歌颂了大自然，并且与来自亚洲和巴西石灰岩细部设计结合在一起。艺术的重制向自然致敬。

大堂里，一面巨大的意大利树皮板（来自瓦努阿图岛）迎接着宾客的到来。来自中国、印度尼西亚、菲律宾、南非、摩洛哥和巴西本土的艺术品重塑了整个空间，实现了建筑师的文化融合策略。进入度假村，人们仿佛到了世界各地，在阿拉戈斯体验到变化万千的世界文化。在2,749平方米的建设中，水元素起到了画龙点睛的作用，凸显了该州"水天堂"的名号。其中一个主要的倒影池群以粗糙的石头与水形成了平衡，让岛屿变得柔和。

柯诺亚度假村的一切都很明晰，处处充满惊喜，重塑了每一个角落。物品在自然状态下演化（如生成铁锈），实现了项目的人性尺度。被改造成装饰花瓶的来自铸铁厂的旧水壶、黄铜淋浴头（由建筑师设计）、氧化铁水龙头、旧制糖机改造成的雕塑、制成并嵌入建筑结构中的沉睡的铁轨等，建筑师花费大量时间来创造新材质和应用设施。

在水疗中心中，建筑师再次向阿拉戈斯州表达了敬意。他以水的运动为灵感，将水疗中心打造成了一座放松的圣殿，让人们与自然邂逅。一个巨大的游泳池和12个私人游泳池四周都砌上了石头，再现了海洋的蔚蓝，而大型极可意按摩浴缸同时以粗糙的石材为美。

独特而严谨的设计和优秀的细节是柯诺亚度假村的独特感官体验。所有物品都源于手工制作，独一无二，展现了柯诺亚的简约与奢华。项目的所有家具都是特别定做的。沙发、桌子、座椅都与自然紧密相连，同时又为项目添加了理想的现代感。

7. Main building outside view
   主建筑外观

8. Spa entrance in the dry forest plaza
   水疗中心干燥林广场入口

9. Restaurant interior
   餐厅室内

10. Restaurant balcony
    餐厅阳台

11. Winebar internal view
    酒吧室内

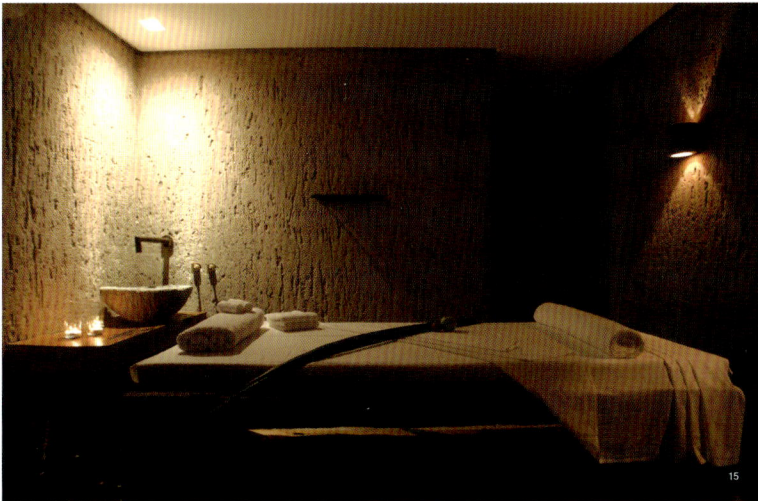

Sheraton Iguazú Resort & Spa

喜来登伊瓜苏水疗度假村

Completion date: 2010 (renovation)
Location: Puerto Iguazú, Argentina
Designer: Estudio Kocourek
Photographer: Sheraton Iguazú Resort & Spa

竣工时间：2010年（翻新）
项目地点：阿根廷，伊瓜苏港市
设计师：Estudio Kocourek设计工作室
摄影师：喜来登伊瓜苏水疗度假村

Sheraton Iguazú Resort & Spa is a complete paradise surrounded by nature. Iguazú, which means 'the great waters' in the Guarani language, is located in the northern part of the province of Misiones, right on the Argentine-Brazilian frontier. Listed as a UNESCO World Natural Heritage Site and voted one of the New 7 Wonders of Nature, the Iguazú Falls and Iguazú National Park's natural splendour and lush vegetation are a must-see. The falls extend over 2,700 metres, with over 260 waterfalls reaching 70 metres high. They can be appreciated from Argentina or Brazil, with the magnificent Devil's Gorge defining

the border between the two countries.

The spacious terrace of the Lobby Bar is the perfect place to connect with nature thanks to its amazing view of the falls and great cocktails, snacks, and pastries. The three-floor resort offers 176 spacious guest rooms with views of the Iguazú Falls or the rainforest. With a private balcony and large windows, you'll feel even closer to nature. All rooms are fully appointed and equipped with individually controlled air conditioning, high speed Internet access, the exclusive Sheraton Sweet Sleeper™ Bed, and much more.

The spectacular new SEDA Spa features a variety of treatment options in five indoor boxes, four outdoor tents, one internal box for couples, and a Vichy shower. Additionally, there is a spacious Zen relaxing room, a salon, a sauna, a steam room, and a large hot water whirlpool for therapeutic treatments.

Additional recreational facilities include a modern, fully-equipped gym, a children's pool, and an adult pool with a pool bar. A tennis court, complete with lights, is also available. The business centre meets the needs of business travellers, providing comfort and privacy in modern surroundings. Facilities include a

meeting room, two works stations for notebooks, and four stations equipped with PCs, photocopier, broadband, and wireless high speed Internet access.

1. Resort exterior and Iguazu falls
   度假村外观与伊瓜苏瀑布
2. Oasis terrace banquet
   绿洲露台宴会场所
3. Front desk area
   前台
4. Lobby bar with view of the falls
   可以欣赏瀑布美景的大堂酒吧
5. Adelantado ballroom banquet
   贵族宴会厅
6. Resort plan
   度假村平面图

1. Access
2. Business centre
3. Deposit equip
4. Reception
5. concierge
6. Lobby bar
7. Sheraton link
8. Terrace
9. Salon

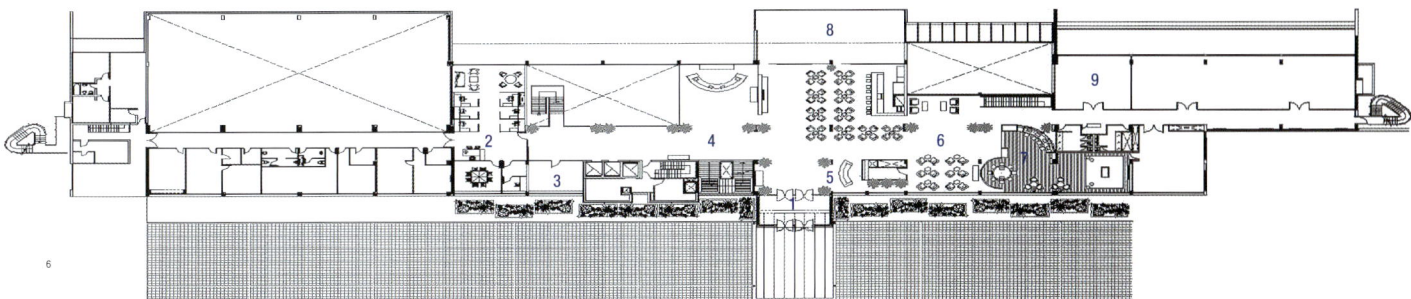

1. 通道
2. 商务中心
3. 存储设备间
4. 接待处
5. 门房
6. 大堂酒吧
7. 喜来登网络部
8. 露台
9. 沙龙

喜来登伊瓜苏水疗度假村完全是一个自然美景环绕的度假天堂。伊瓜苏在瓜拉尼语中意思是"伟大的水域"，位于米西奥内斯省北部，正好在阿根廷与巴西接壤处。被列为世界自然遗产和世界新七大自然奇观的伊瓜苏瀑布和伊瓜苏国家公园内的自然美景和繁茂的植物是这里的必看之处。瀑布遍及2700米，其中有260米的瀑布高达70米。阿根廷和巴西以魔鬼峡谷为边界，从两国境内都可以观赏到此壮观的景色。

大堂吧宽敞的阳台是与大自然接触的完美之地，因为从这里可以欣赏瀑布的惊人美景，还可以品尝鸡尾酒、小吃及甜点。这个三层楼的度假村建筑提供176间宽敞的客房，可以欣赏伊瓜苏瀑布和雨林的迷人景色。客房内的私人阳台和大大的窗户会让你感觉与大自然更加亲近。所有的客房都设配齐全，配有独立控制的空调、高速上网、喜来登专属甜梦睡床等。

新建的SEDA水疗中心装饰迷人，设有5间室内治疗间，4间户外棚区治疗间，一间室内双人治疗室和一间维希淋浴间，提供一系列特色的治疗。另外，还有一间宽敞的禅宗休息室、沙龙、桑拿浴室、蒸汽室和一个用于水疗治疗的热水漩涡浴缸。

其他休闲娱乐设施有现代设备齐全的健身房、儿童泳池和带泳池吧的成人泳池。还有配有灯光的网球场。商务中心能够满足商务旅行者的需求，以现代的装饰提供舒适与私密的空间。服务设施包括一间会议室，两个工作站可供笔记本工作，还有4个工作站配有台式电脑、影印机、宽带和无线高速上网。

7. Seda pool and spa exterior treatment tents
   Seda泳池与水疗户外治疗篷区
8. Outdoor pool
   户外游泳池
9. Seda pool and spa reception
   Seda泳池与水疗接待处

10. Seda pool, spa Zen room and exterior treatment tents
    Seda泳池，水疗禅室与户外治疗篷区
11. Seda pool and spa massage cabin
    Seda泳池与水疗按摩间
12. Seda pool and spa couples whirlpool
    Seda泳池与水疗双人漩涡浴缸
13. Suite living room
    套房客厅
14. Premier suite
    首席套房
15. Falls view room
    瀑布观景客房

Bora Bora Pearl Beach Resort & Spa

波拉波拉岛珍珠滩水疗度假村

Completion date: 2011 (renovation)
Location: Bora Bora, French Polynesia
Designer: Pierre Jean Picard
Photographer: Bora Bora Pearl Beach Resort & Spa

竣工时间：2011年（翻新）
项目地点：法属波利尼西亚，波拉波拉岛
设计师：皮埃尔·珍·皮卡德
摄影师：波拉波拉岛珍珠滩水疗度假村

Bora Bora Pearl Beach Resort & Spa is located on Motu Tevairoa facing the island of Bora Bora. It offers a place to luxuriate in the peace and tranquility of the magical surroundings with unsurpassed views on the world renowned Mount Otemanu.

This traditional Polynesian style hotel offers 50 overwater bungalows, 10 beach suites and 20 garden pool suites. The 50 overwater bungalows are 53 square metres each. All offer a very spacious bathroom area with both a separate shower and a full size bath tub. The overwater bungalows have glass tables to view the aquatic life of the lagoon, a large sundeck and sitting area with direct access to the water. Among the

50 overwater bungalows, 30 are classified as premium due to their view. The beach suites offer a lounge area facing the beach and a private enclosed garden with Jacuzzi. The Garden & Premium Pool Suites feature a private tropical garden, a plunge pool and a spacious covered sitting & resting area.

The Bora Bora Pearl Beach Resort has three restaurants, two bars, one boutique, an on-site diving centre — the Top Dive centre, a fresh water swimming pool, one floodlit tennis court, mini golf, table tennis, volley ball court and bacci ball court, billiards and a wide variety of activities and excursions. The large fresh water swimming pool of Bora Bora (485 sqm) is situated at the base of a water cascade with a Jacuzzi located beside the centre of the pool. The pool is just next to the beach and is surrounded by decked and tiled areas with lounge chairs.

The Manea Spa covers an area of 675sqm and combines traditional Tahitian architecture with state of the art equipments and facilities. There are two double and a single massage rooms, a Royal Suite, a double Vichy shower, a beauty & tattoo salon, two saunas, two steam bath and a fully equipped fitness room.

波拉波拉岛珍珠滩水疗度假村坐落于默图泰瓦尔，面朝波拉波拉岛。该度假村是享受和平与宁静之地，周围环境充满神奇的色彩，还可以在这里欣赏到世界闻名的欧特马努山无以伦比的美景。

这个传统的波利尼西亚风格的酒店提供50间水上屋、10套海滨套房和20套花园泳池套房。50间水上屋每间面积达53平方米，并且都提供宽敞的浴室，浴室中有独立的淋浴和全尺寸的浴缸。水上屋内设有玻璃桌子，从这里可以观赏环礁湖的水下生物，还有一个大阳台和与海水相接的休息区。50间水上屋中有30间为高级水上屋，从这里可以观赏美丽的风景。海滨套房提供面向海滩的休闲区和一个带极可意水流按摩浴缸的私人封闭花园。花园和高级泳池套房设有私人热带花园、跌水潭和宽阔的带篷休息区。

波拉波拉岛珍珠滩水疗度假村有3间餐厅，两间酒吧，一间精品店，一个现场潜水中心——顶级潜水中心，一个淡水游泳池，一个泛光灯照明的网球场，迷你高尔夫场，乒乓球场，排球场和巴捷尔球场，台球区和其他一系列活动和远足。波拉波拉淡水游泳池（485平方米）位于一个小瀑布的底部，在游泳池中心边上有一个极可意水流按摩浴缸。游泳池就位于海滩旁，周围是甲板躺椅区。

马尼亚水疗中心覆盖面积达675平方米，结合了传统的塔希提建筑风格与先进设备和服务设施。该水疗中心有两间双人按摩室和一间单人按摩室，一间皇家套房，一间双人维西淋浴室，一间美容与文身沙龙，两间蒸汽浴室和一间设备完善的健身室。

1. Reception
   Tevairoa Restaurant
   Taurearea Bar
   Movie theatre
   Conference centre
   Wedding chapel
2. Miki Miki Restaurant & Ambrosia Restaurant
   Miki Miki Bar
3. Swimming pool + Jacuzzi
4. Diving centre
5. Tennis court & volley ball
6. Manea Spa & Fitness
7. Garden Pool Suites
8. Beach Suites with Jacuzzi
9. Overwater Bungalows
10. Premium Overwater Bungalows
11. Way to heliport
12. Mini golf
13. To' A Nui Coral nursery
14. Arrival and departure pontoon

1. 接待处
   Tevairoa餐厅
   Taurearea酒吧
   电影院
   会议中心
   婚礼教堂
2. Miki Miki餐厅与安布罗西亚餐厅
   Miki Miki酒吧
3. 游泳池与极可意水流按摩浴缸
4. 潜水中心
5. 网球场与排球场
6. 摩尼亚水疗中心与健身中心
7. 花园泳池套房
8. 带极可意水流按摩浴缸的海滨套房
9. 水上屋
10. 高级水上屋
11. 通往直升机场
12. 迷你高尔夫球场
13. To' A Nui Coral托儿所
14. 到达与离开的浮桥

1. Aerial view of the overwater bungalows
   水上屋鸟瞰图
2. Overwater pontoon
   海上浮桥
3. Beach view
   海滩景色
4. Beach view at night
   海滨夜景
5. Resort plan
   度假村平面图
6. Manea Spa entrance
   马尼亚水疗中心入口
7. Tevairoa Restaurant
   泰瓦偌餐厅
8. Lobby
   大堂

9. Manea Spa corridor
   马尼亚水疗中心通道
10、11. Manea Spa Vichy shower
   马尼亚水疗中心维西淋浴
12. Manea Spa relaxation
   马尼亚水疗中心休息区
13. Manea Spa massage
   马尼亚水疗中心按摩区

14. Manea Spa Vichy shower
   马尼亚水疗中心维西淋浴
15. Manea Spa massage double
   马尼亚水疗中心双人按摩室
16. Beach bungalow
   海滨小屋
17. Garden Pool Suite
   花园泳池套房
18. Aerial view of the Garden Pool Suite
   花园泳池套房鸟瞰图

19. Beach suite bedroom
    海滨套房卧室
20. Overwater suite bathroom
    水上套房盥洗室
21、22. Overwater bungalow
    水上屋

Moorea Pearl Resort & Spa

茉莉雅珍珠水疗度假村

Completion date: 2007 (renovation)
Location: Moorea, French Polynesia
Designer: Pierre Lacombe
Photographer: Moorea Pearl Resort & Spa
Area: 30, 351sqm

竣工时间：2007年（翻新）
项目地点：法属波利尼西亚，茉莉雅岛
设计师：皮埃尔·拉科姆
摄影师：茉莉雅珍珠水疗度假村
项目面积：30, 351平方米

Moorea Pearl Resort & Spa is located at a short drive from the arrival points on the island and two miles from the magnificent Cook's Bay. The Moorea Pearl Resort & Spa features three stone tiki sculptures (mythical figures of Polynesia) that honour the rich Polynesian culture and traditions. The sculptures guard over the lobby area, the garden and the pool.

This charming resort is a perfect blend of luxury and serenity. This traditional Polynesian style resort offers 94 rooms and bungalows. It features 28 Overwater Bungalows, 8 Beach Bungalows, 28 Garden Bungalows and 30 Garden Rooms & Family Rooms.

Moorea Pearl Resort & Spa offers garden bungalows with private plunge pool, but guests may also enjoy the clear water of the infinity swimming pool, the largest one on the island. In addition, all bungalows offer a direct access to the lagoon. Premium overwater bungalows set right above the reef drop off allowing a fabulous snorkelling.

The hotel proposes 2 restaurants, Le Mahanai and the prestigious Gourmet restaurant Le Matiehani. The terrace of the bar Autera'a is also a perfect place for an afternoon snack. Other facilities available include 1 bar, 2 meeting rooms, 1 boutique and a wide range of land or water activities and excursions. The hotel has an on site diving centre, a spa and Polynesian tattoos. Manea Spa is nestled in the heart of the resort. The Spa has been conceived and built in the pure tradition of Polynesian well-being and traditions. This Manea Spa features 1 hammam, 3 massage rooms, 1 lobby, 1 room with shower with 4 jets, 1 room with Vichy shower, 1 special pool for aquatic massage.

1. Reception
   Hall
   Boutique
2. Conference centre
3. Mahanai Restaurant
4. Autera' a Bar
5. Manea Spa
6. Garden View Room and Garden View Duplex
7. Garden Pool Bungalows
8. Beach Bungalows
   Diving centre
9. Premium Beach Bungalows
10. Overwater Bungalows
11. Premium Overwater Bungalows
12. Swimming pool

1. 接待处
   礼堂
   精品店
2. 会议中心
3. Mahanai餐厅
4. Autera' a酒吧
5. 摩尼亚水疗中心
6. 花园景观客房与花园景观奢华客房
7. 花园泳池小屋
8. 海滨小屋
   潜水中心
9. 高级海滨小屋
10. 水上屋
11. 高级水上屋
12. 游泳池

1. Overwater bungalows
   水上屋

2. Garden Suite pool
   花园套房泳池

3. Overwater bungalow terrace
   水上屋露台

4. Resort plan
   度假村平面图

5. Duplex Garden room
   双层花园客房

6. Premium Garden Pool bungalow
   顶级花园泳池别墅

7. Lobby
   大堂

8、9. Beach bungalow bedroom
   海滨别墅卧室

茉莉雅岛珍珠水疗度假村距离岛上到达地点只需短暂的车程，距离美丽的库克湾只有两英里。茉莉雅岛珍珠水疗度假村有三座特色的石刻提基雕像(波利尼西亚神话人物)，象征着丰富的波利尼西亚文化与传统。这几座雕像守护着大堂区、花园和泳池区。

这个迷人的度假村将奢华与宁静完美地融合在一起。度假村为传统的波利尼西亚风格，提供94间客房与小屋。其中有28间水上屋，8间海滨小屋，28间花园小屋和30间花园客房与家庭客房。茉莉雅岛珍珠水疗度假村提供的花园小屋带有私人跌水潭，但是客人们或许还是会喜欢享受岛上最大的无边泳池中清澈的海水。而且，所有的小屋都可以直达环礁湖。高级水上屋建在礁石稀少区的上方，能够进行难以置信的潜水活动。

酒店有2间餐厅：Le Mahanai餐厅和著名的Le Matiehani美食餐厅。Autera'a酒吧的阳台是享受下午小吃的完美之地。提供的其他服务设施有：一间酒吧，两间会议室，一间精品店和一系列陆地与水上活动和远足。酒店有一个现场潜水中心，一个水疗中心和波利尼西亚文身中心。马尼亚水疗中心隐于度假村的中心，是按照纯粹的波利尼西亚美容与传统而构思和建造的。马尼亚水疗中心设有一间哈曼浴室，3间按摩室，一个大堂，一间带有四个喷射头的淋浴室，一间维西淋浴室和一个为水疗按摩准备的特别游泳池。

10. Garden Pool bungalow
   花园泳池别墅
11. Beach bungalow
   海滨别墅
12. Overwater bungalow
   水上屋
13. Garden View Duplex
   双层花园景观客房

Grand Hyatt Kauai Resort & Spa

考艾岛君悦水疗度假村

Completion date: 2011(renovation)
Location: Kauai, Hawaii
Designer: Wimberly Allison Tong & Goo
Photographer: Grand Hyatt Kauai Resort & Spa
Area: 202,342sqm

竣工时间：2011年（翻新）
项目地点：夏威夷，考艾岛
设计师：Wimberly Allison Tong & Goo建筑事务所
摄影师：考艾岛君悦水疗度假村
项目面积：202, 342平方米

Set on a soothing white sand beach, the Grand Hyatt Kauai Resort & Spa unfolds among lush gardens and pristine lagoons. Classic Hawaiian style architecture featuring open courtyards and walkways keeps the elegant atmosphere informal and reflective of Hawaii's climate and culture.

Strongly committed to environmental conservation and preservation of natural resources, this luxury resort has it all from a veritable water playground with a lava-rock river pool, waterslide, and saltwater lagoon to 18-holes of championship golf at the Poipu Bay Resort. Sample the delicious flavours of the island

at the resort's six restaurants. And for a Hawaiian spa experience like no other, Anara Spa offers treatments in outdoor garden bungalows using island-fresh botanical essences blended with traditional healing customs.

Grand Hyatt Kauai Resort & Spa has just finished a thorough renovation of their 37 suites. In the resort's legendary Hawaiian classic style, suites offer all the comforts of home with a graceful elegance. Creating an exclusive retreat for guests, tastefully appointed suites offer refined furnishings and design elements from land and sea. Hardwood flooring, artistic floral rugs, and furniture with timeless, clean lines pay homage to the iconic island home as the decor echoes the multi-cultural influences of the island. The comfortable, modern lifestyle blends seamlessly with Hawaiian touches as colours reflect the blue of the sea, white of the sand, and ochre of the land.

Garden, Ocean and Deluxe Suites range between 93-167 square metres and feature a dining area with a granite top wet bar, spacious living areas, and a separate bedroom. The 223 plus square metres Presidential Suites offer a full kitchen and dining room, a living room, entertainment area with surround

1. Poipu Guestroom Wing
2. Shipwreck Wing
3. Freshwater pool
4. Shipwreck Beach Lagoon
5. Saltwater Lagoons
6. Grand Club Lounge
7. Waterslide
8. Camp Hyatt
9. Poipu Bay Golf Course
10. Tennis Courts
11. Logo Shop/Bicycle Centre
12. Anara Spa and Salon
13. Main Lobby
14. Main Porte Cochere
15. Keoneloa Bay
16. Resort Shopping on Both Sides of Lobby
17. Indoor Meeting Space
18. Grand Garden
19. Group Porte Cochere
20. Ilima Garden
21. Poipu Garden
22. Resolution Garden
23. Shipwreck Beach Lagoon
24. Cabana Beach
25. Wedding Gazebo Ilima Lawn
26. Wedding Gazebo Regency Lawn
27. Seaview Terrace
28. Captain's Bar
29. Dondero's Restaurant
30. Stevenson's Library
31. The Dock Restaurant
32. Tidepools Restaurant/Lounge
33. Ilima Terrace
34. Shipwreck Bar

1. 波普客房区
2. 西普莱克客房区
3. 淡水游泳池
4. 船难海滩环礁湖
5. 咸水环礁湖
6. 大俱乐部休闲吧
7. 水滑梯
8. 君悦露营区
9. 帕伊普海湾高尔夫球场
10. 网球场
11. 标识商店、自行车中心
12. 阿娜拉水疗中心与沙龙
13. 大厅
14. 主车辆门道
15. 奇欧尼罗海湾
16. 大厅两侧度假村商店
17. 室内会议空间
18. 大花园
19. 团体车辆门道
20. 伊利玛花园
21. 帕伊普花园
22. 决议花园
23. 船难海滩环礁湖
24. 海滨小屋
25. 婚礼亭伊利玛草坪
26. 婚礼亭摄政草坪
27. 海景露台
28. 船长酒吧
29. 唐德罗餐厅
30. 斯蒂文森图书室
31. 码头餐厅
32. 蓄潮池餐厅
33. 伊利玛露台
34. 海难酒吧

1、2. Resort golf course
度假村高尔夫球场

3. Resort plan
度假村平面图

4. Resort water view
度假村海景

5. Resort exterior sunset
日落时分度假村外景

6. Spa Vichy shower
水疗中心维西淋浴

7. Suite detail
套房细部

8. Guestroom
客房

9. Suite living room
套房客厅

10. Suite bedroom
套房卧室

sound, two bathrooms, and a separate bedroom. All suites feature spacious decks, a bath with a whirlpool tub, walk-in rain style shower, double marble sink vanities and all the amenities of home.

All rooms are designed to maximise guests' comfort and convenience while advancing the resort's conservation efforts. Adding to the resort's list of environmental initiatives the renovation included Toto low-flow toilets, low VOC paints, recycled yarn carpet, and energy-saving ceiling fans. Clean lines and crisp, fresh colours combine to create a relaxed elegance in a distinct Hawaiian classic style, making guests feel at home.

考艾岛君悦水疗度假村位于令人舒缓的白色沙滩上，掩映于繁茂的花园与原始环礁湖中。经典的夏威夷风格建筑，开放式的庭院与通道设计营造了一种反映夏威夷气候与文化的典雅氛围。

坚定不移地致力于保护环境和自然资源，该奢华度假村拥有一个真正的水上游乐场，游乐场里有火山石淡水泳池、水滑梯和海水环礁湖，还拥有一个位于波普湾的18洞高尔夫锦标赛球场。度假村有6间餐厅可以提供岛上各种风味的美食。阿娜拉水疗中心提供与众不同的夏威夷水疗体验，这里有户外花园小屋治疗室，使用岛上新鲜植物精华，结合传统治疗手法进行治疗。

考艾岛君悦水疗度假村刚结束了对37间套房的翻新。在度假村传奇的夏威夷古典风格中，典雅的套房提供一种如家里一般的舒适感。为客人创造一个专属的避世之地，设计典雅的套房提供来自陆地与海洋的精美装饰和设计元素。装饰风格受岛上多元文化的影响，实木地板，艺术花地毯，经典的家具，优美的线条相结合创造标志性的岛上家园。现代舒适的生活方式与夏威夷色彩，如海水蓝，沙滩白，还有陆地赭色，天衣无缝地结合在一起。

花园套房、海洋套房和奢华套房，面积从93平方米到167平方米，有带花岗岩顶级吧台的餐厅，宽敞的起居室和独立的卧室。223平方米的总统套房提供设备齐全的厨房与餐厅、起居室、带环绕立体声的娱乐区、两间浴室和一间独立卧室。所有的套房都配有宽敞的阳台，带漩涡浴缸的浴室，步入式雨淋浴室，双大理石水槽和所有家里应有的设施。

所有客房的设计都最大程度地满足了客人对舒适与便捷的需求，同时又极大地致力于环境保护。度假村在翻新过程中加入了一系列新的环保措施，包括东陶节水厕所，低VOC涂料，可循环使用的纱线地毯，还有节能天花板风扇。优美的曲线结合清新明亮的色彩，营造出独特的夏威夷古典风格中轻松优雅的氛围，使客人感觉如在家中般轻松自在。

Castaways Resort & Spa

卡斯特维斯水疗度假村

Completion date: 2010
Location: Queensland, Australia
Designer: Gordon Beath, Edge Architecture
Photographer: Melanie Ryan

竣工时间：2010年
项目地点：澳大利亚，昆士兰
设计师：戈登·比斯，艾吉建筑事务所
摄影师：梅勒妮·赖安

This simple building with strong repetitive forms became a symbolic place within the Mission Beach community. The hotel suffered minor amendments during this time. The very limited service, lack of physical relationship with the beachfront and lack of contact with members of the community became key elements to overcome upon the arrival of new owners. The creation of the revitalised resort experience commenced with a strong engagement between the building and the foreshore parklands. Activities from arrival, dining, relaxing, and swimming are all linked intrinsically to the spectacular beachfront and island

views.

The existing gardens were stripped back and enhanced to offer sightlines into the building form, a new vehicle and pedestrian entry established to welcome guests and the local community alike. The redesign emphasises the pedestrian linkages between the surrounding streets, the esplanade and the resort. The use of vertical timber panels and screens rhythmically complement the existing strong gabled roof form providing a richly articulated and visually dynamic exterior towards the beachfront and parklands and a comfortable scale for pedestrians.

Multiple outdoor timber decking areas surround the two new pools, with the lap pool promoting active recreation, and the smaller pool offering more relaxed ambience adjacent the beverage bar. The decking areas offer a range of spatial and private experiences, to suit the varied taste of guests. Informal areas have been created using visually permeable timber screens to provide a sense of privacy whilst allowing the individual to participate in the surrounding activities, i.e. swimming, sun bathing, dining or beverages by the bar.

The internal spaces were reconfigured to accommodate

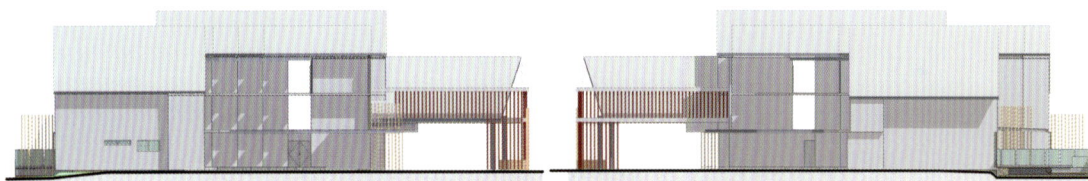

1. Building exterior
   建筑外观
2. Elevation
   立面图
3. Section
   剖面图

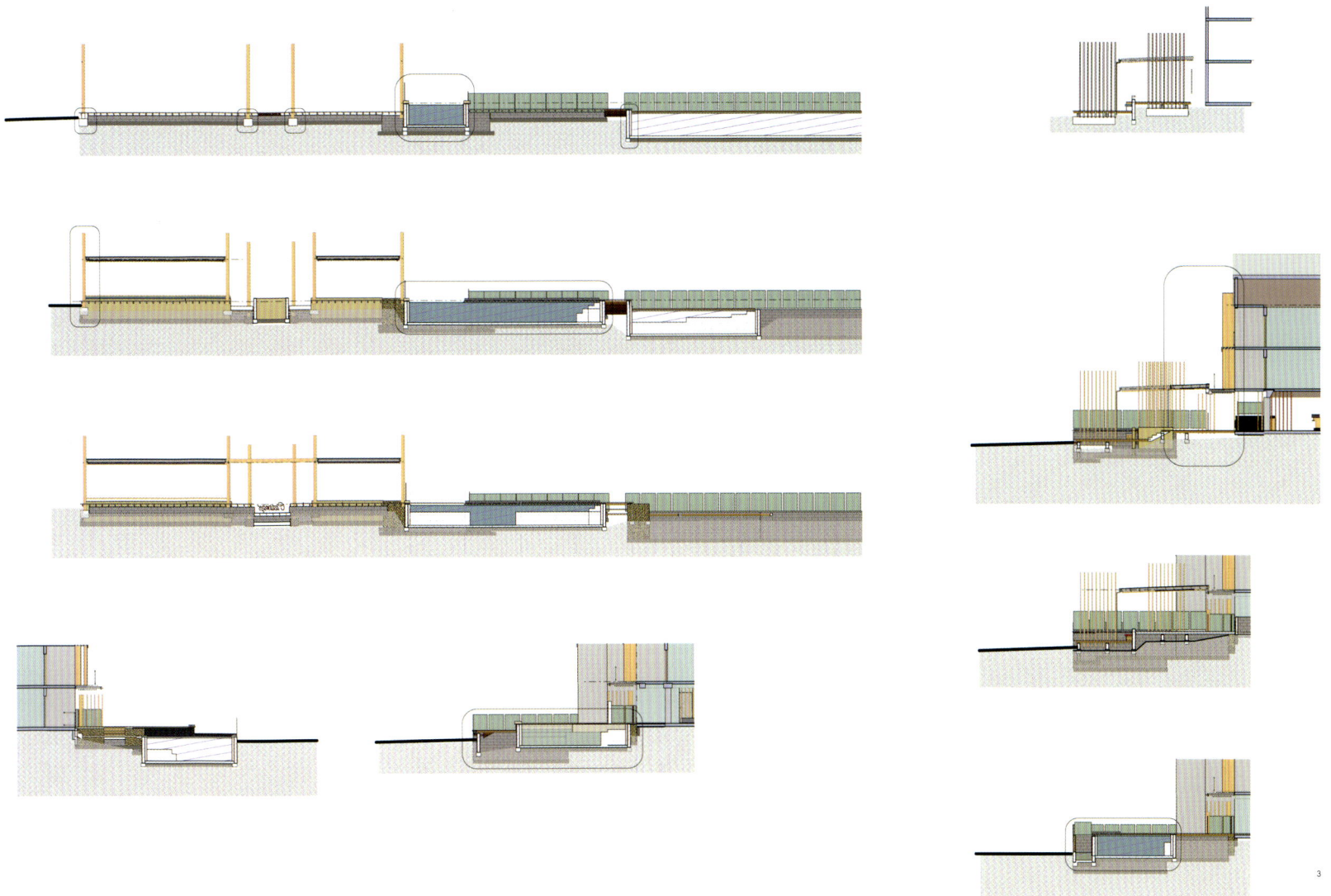

reception-lobby seating space, beverage bar, display kitchen, lounge areas and bathrooms. Interior design is visually stimulating, the simple palette of materials linking both interior spaces and outdoor areas promoting through breezes, natural light, visual lines and vistas. Blackbutt timber dominates throughout and finished with respect to its location, transitioning from naturally weathered at the building edges, stained outdoor elements and polished internally. Timber elements are enhanced with bluestone mosaic tiles and complementary coloured veneers and paint. The black tiled bathrooms providing a welcome respite from the tropical glare.

The spa facilities include private treatment rooms open to the beach and specialist treatment rooms within a quaint beachside cottage, including vichy shower.

这个简单的建筑是由同一建筑形式的重复构成的，是密逊海滩社区一个象征性的地方。酒店经历了小规模的翻新。非常有限的服务、缺少与海滨和社区成员的联系是新业主需要克服的主要因素。创造度假村的重新复兴首先从建立酒店建筑与海滨公园的关系开始。从到达度假村到就餐、休闲放松，再到游泳等一系列活动都要与海滨和岛上的壮观景色紧密的联系在一起。

为了不遮挡建筑物前的视线，现有的花园被移除，进而建立一个新的车辆与行人入口，来迎接客人和当地社区人员。重新设计非常重视周围街道、游憩场和度假村之间步行通道的连接。有节奏地使用竖立的木板和屏风，与已有的三角形屋顶相辅相成，形成一个视觉上清晰的动态外观，通向海滩与海滨公园，对行人来说是一个令人舒适的规模。

两个新建的游泳池旁有大量的户外木制甲板区，小型健身游泳池以运动康乐为主，小一点的游泳池与提供饮料的休息吧相连接，提供更加休闲放松的氛围。甲板区提供一系列具有空间感和私密感的体验活动，适合不同品位的客人。利用可透视木制屏风建立的非正式区域提供一种私密感，同时每个人都可以参与周围的各种活动，如游泳、日光浴、就餐或在休息吧中喝点饮料。

室内空间经过重新装饰的地方有接待处大堂休息区、饮料休息吧、开放厨房、休闲厅和浴室。室内的设计在视觉上令人振奋，颜色简单的材料将室内空间与室外区域相连接，并且更利于享受微风、自然光、与户外景色。黑基木的使用贯穿整个建筑始终，并且根据不同的位置来做装饰，建筑边缘、户外染色元素和内部磨光的黑基木已经自然褪色了。木制元素与青石马赛克瓷砖和当代彩色胶合板和涂料搭配使用。铺有黑色瓷砖的浴室为客人在热带的炫光中提供一个喘息之地。

水疗中心设施有面向海滩的私人治疗室，还有位于海滨古雅小屋里的特别治疗室，里面有维希淋浴室。

4. Restaurant outdoor
   户外餐厅
5. Bar outdoor
   户外酒吧
6. Building entrance
   建筑入口
7. Section
   剖面图
8. Lobby lounge
   大堂吧
9. Bibesia beach club
   贝西亚海滨俱乐部

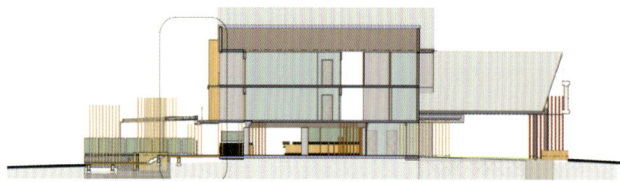

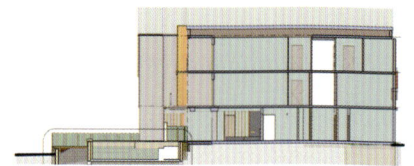

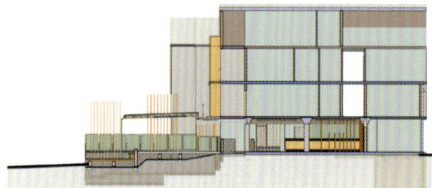

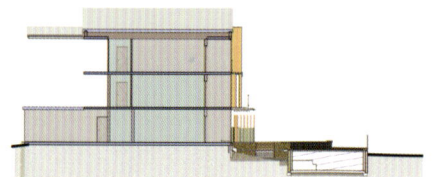

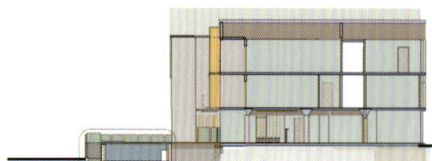

10. Spa relaxation area
    水疗中心休息区
11. Spa entrance
    水疗中心入口
12. Spa relaxation area
    水疗中心休息区
13. Spa massage room
    水疗中心按摩室
14. First floor plan
    一层平面图

1. Dining
2. Pool terrace
3. Bar
4. Store
5. Pool
6. Balcony
7. Guest room
8. Lap pool deck
9. Admin

1. 餐厅
2. 泳池露台
3. 酒吧
4. 储藏室
5. 游泳池
6. 阳台
7. 客房
8. 小型健身游泳池甲板
9. 行政管理处

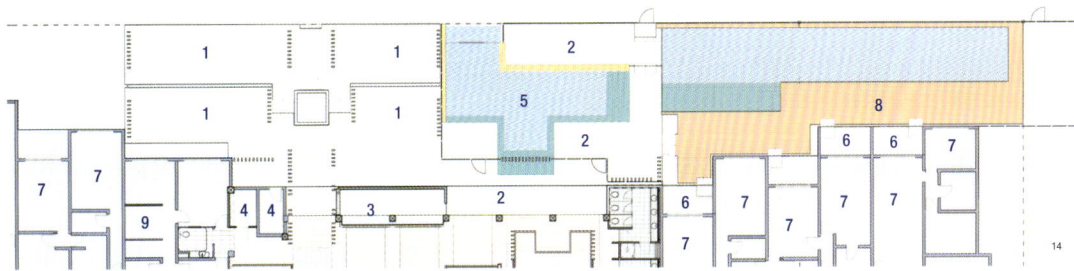

Index 索引